Creo Manufacturing 4.0
Black Book
(Colored)

By
Gaurav Verma
Matt Weber
(CADCAMCAE Works)

Edited by
Kristen

ISBN # 978-1-988722-15-3

NOTICE TO THE READER

Publisher does not warrant or guarantee any of the products described in the text or perform any independent analysis in connection with any of the product information contained in the text. Publisher does not assume, and expressly disclaims, any obligation to obtain and include information other than that provided to it by the manufacturer.

The reader is expressly warned to consider and adopt all safety precautions that might be indicated by the activities herein and to avoid all potential hazards. By following the instructions contained herein, the reader willingly assumes all risks in connection with such instructions.

The Publisher makes no representation or warranties of any kind, including but not limited to, the warranties of fitness for a particular purpose or merchantability, nor are any such representations implied with respect to the material set forth herein, and the publisher takes no responsibility with respect to such material. The publisher shall not be liable for any special, consequential, or exemplary damages resulting, in whole or part, from the reader's use of, or reliance upon, this material.

DEDICATION

To teachers, who make it possible to disseminate knowledge
to enlighten the young and curious minds
of our future generations

To students, who are the future of the world

THANKS

To my friends and colleagues

To my family for their love and support

Training and Consultant Services

At CADCAMCAEWORKS, we provides effective and affordable one to one online training on various software packages in Computer Aided Design(CAD), Computer Aided Manufacturing(CAM), Computer Aided Engineering (CAE), Computer programming languages(C/C++, Java, .NET, Android, Javascript, HTML and so on). The training is delivered through remote access to your system and voice chat via Internet at any time, any place, and at any pace to individuals, groups, students of colleges/universities, and CAD/CAM/CAE training centers. The main features of this program are:

Training as per your need

Highly experienced Engineers and Technician conduct the classes on the software applications used in the industries. The methodology adopted to teach the software is totally practical based, so that the learner can adapt to the design and development industries in almost no time. The efforts are to make the training process cost effective and time saving while you have the comfort of your time and place, thereby relieving you from the hassles of traveling to training centers or rearranging your time table.

Software Packages on which we provide basic and advanced training are:

CAD/CAM/CAE: CATIA, Creo Parametric, Creo Direct, SolidWorks, Autodesk Inventor, Solid Edge, UG NX, AutoCAD, AutoCAD LT, EdgeCAM, MasterCAM, SolidCAM, DelCAM, BOBCAM, UG NX Manufacturing, UG Mold Wizard, UG Progressive Die, UG Die Design, SolidWorks Mold, Creo Manufacturing, Creo Expert Machinist, NX Nastran, Hypermesh, SolidWorks Simulation, Autodesk Simulation Mechanical, Creo Simulate, Gambit, ANSYS and many others.

Computer Programming Languages: C++, VB.NET, HTML, Android, Javascript and so on.

Game Designing: Unity.

Civil Engineering: AutoCAD MEP, Revit Structure, Revit Architecture, AutoCAD Map 3D and so on.

We also provide consultant services for Design and development on the above mentioned software packages

For more information you can mail us at:
cadcamcaeworks@gmail.com

Table of Contents

Chapter 3 : Tool Setting

Chapter 4 : Machine Setting

Chapter 6 : Creating toolpaths and Output

Chapter 7 : CMM Introduction

Chapter 8 : Project

Preface

Creo Parametric 4.0 is a parametric, feature-based solid modeling tool that not only unites the three-dimensional (3D) parametric features with two-dimensional (2D) tools, but also addresses every design-through-manufacturing process. The continuous enhancements in the software has made it a complete PLM software. The software is capable of performing analysis with an ease. Its compatibility with CAM software is remarkable. Based mainly on the user feedback, this solid modeling tool is remarkably user-friendly as compared to its previous versions and it allows you to be productive from day one.

To support education, PTC gives free educational version of the Creo to students with a license of 1 year. You can get your free version of software by visiting the following link:

https://www.ptc.com/en/academic-program/products/free-software

The **Creo Manufacturing 4.0 Black Book** is the first edition of our series on Creo Parametric Manufacturing Apps. The book is crafted to help professionals as well as students in creating basic CAM programs using Creo Parametric. The book follows a step by step methodology. In this book, we have tried to give real-world examples with real challenges in manufacturing. We have tried to reduce the gap between university use of Creo Parametric and industrial use of Creo Parametric. The book starts with basics of machining and ends at advanced operations in milling. Some of the salient features of this book are :

In-Depth explanation of concepts

Every new topic of this book starts with the explanation of the basic concepts. In this way, the user becomes capable of relating the things with real world.

Topics Covered

Every chapter starts with a list of topics being covered in that chapter. In this way, the user can easy find the topic of his/her interest easily.

Instruction through illustration

The instructions to perform any action are provided by maximum number of illustrations so that the user can perform the actions discussed in the book easily and effectively. There are about 1200 small and large illustrations that make the learning process effective.

Tutorial point of view

At the end of concept's explanation, the tutorial make the understanding of users firm and long lasting. Almost each chapter of the book has tutorials that are real world projects. Moreover most of the tools in this book are discussed in the form of tutorials.

Project

Free projects and exercises are provided to students for practicing.

For Faculty

If you are a faculty member, then you can ask for video tutorials on any of the topic, exercise, tutorial, or concept.

Formatting Conventions Used in the Text

All the key terms like name of button, tool, drop-down etc. are kept bold.

Free Resources

Link to the resources used in this book are provided to the users via email. To get the resources, mail us at **cadcamcaeworks@gmail.com** with your contact information. With your contact record with us, you will be provided latest updates and informations regarding various technologies. The format to write us mail for resources is as follows:

Subject of E-mail as **Application for resources of _____ book.**
Also, given your information like
Name:
Course pursuing/Profession:
Contact Address:
E-mail ID:

Note: We respect your privacy and value it. If you do not want to give your personal informations then you can ask for resources without giving your information.

About Authors

The author of this book, Matt Weber, has written many books on CAD/ CAM/CAE available already in market. **SolidWorks Simulation Black Books** are one of the most selling books in SolidWorks Simulation field. The author has hands on experience on almost all the CAD/CAM/CAE packages. Besides that he is a good person in his real life, helping nature for everyone. If you have any query/doubt in any CAD/CAM/CAE package, then you can contact the author by writing at cadcamcaeworks@gmail.com

The author of this book, Gaurav Verma, has authored and assisted in more than 10 titles in CAD/CAM/CAE which are already available in market. He has authored **AutoCAD Electrical Black Books** which are available in both **English** and **Russian** language. He has also authored **Creo Parametric 4.0 Black Book** which covers all designing modules of Creo Parametric. He has provided consultant services to many industries in US, Greece, Canada, and UK.

For Any query or suggestion

If you have any query or suggestion, please let us know by mailing us on *cadcamcaeworks@gmail.com*. Your valuable constructive suggestions will be incorporated in our books and your name will be addressed in special thanks area of our books on your confirmation.

Chapter 1

Beginning with Manufacturing

Topics Covered

The major topics covered in this chapter are:

- *Introduction to Manufacturing*
- *Types of Machines*
- *NC Machines*
- *Applications of CAM*
- *Starting Creo Parametric Manufacturing*
- *Creo Parametric Manufacturing User Interface*
- *Common Operations like Save, Close, Send, and so on*
- *Advantages and Disadvantages of CAM*

INTRODUCTION TO MANUFACTURING

Manufacturing is the process of creating a useful product by using a machine, a process, or both. To manufacture a product, there are some steps to be followed:

- Generating model of final product.
- Deciding Workpiece; selection of raw material and shape/size of workpiece depends on the application of the product.
- Forging, Casting or any other pre-machining method for creating layout for final shape.
- Roughing processes.
- Finishing Processes.
- Quality Control.

As the "Generating model of final product" is above all the steps, it is the most important step. One should be very clear about the final product because all the other steps are totally dependent on the first step. The final product of this step can be a drawing or a model created by using any modeling software.

The next step is "Deciding Workpiece". This step is solely dependent on the first step. Our final product defines what should be the raw material and the workpiece shape/size. Here workpiece is the piece of raw material to be used for the next step or process.

The next step is "Forging, Casting or any other process for creating outline of the final shape". The outline created for the final shape is also called Blank in industries. In this step, various machines like Press, Cutter or Molding machines are used for creating the blank. In some cases of Casting, there is no requirement of machining processes. For example, in case of Investment casting there is generally no requirement of machining process. Machining processes can be divided further into two processes:

- Roughing Processes
- Finishing Processes

These processes are the main area of concern in this book. An introduction of these processes is given next.

Roughing Process

Roughing process is the first cycle of machining processes. Generally, roughing process is the removal of large amount of stock material in comparison to finishing process. In a roughing process, the quantity of material removed from the workpiece is more important than the quality of the machining. Generally, there are no close tolerances for roughing processes. So, these processes are relatively cheaper than the finishing processes. In manufacturing industries, there are three principle machining processes called Turning, Milling, and Drilling. In case of roughing process, there can be turning, milling, drilling, combination of any two, or all the processes. Apart from these machining processes, there are various other processes like shaping, planing, broaching, reaming and so on. But these processes are used only in some special cases.

Finishing Process

Finishing process can include all the machining processes discussed in case of roughing processes but in close tolerances. Also, the quality of machining up to the required level is very important. Apart from the above discussed machining processes, there are a few more machining processes like Electric Discharge Machining(EDM), Laser Beam Machining, Electrochemical Machining and so on. These processes are called unconventional machining processes because of their cutting method. In unconventional machining processes, the tool life is much higher than the conventional machining. These machines are discussed next.

TYPES OF MACHINES

There is a specific machine for every type of machining process. For example for turning process, there is a machine like conventional lathe or CNC Turner. Similarly for milling process, there is a machine called Milling machine. Various machines used for different machining operations are discussed next.

Turning Machines

Turning machines is a category of machines used for turning process. In these type of machines, the workpiece is held in a chuck (collet in case of small workpieces). The chuck revolves at a predefined rotational speed. The chuck can revolve in both CW(Clockwise) or CCW(Counter-Clockwise) direction but cannot translate in any direction, refer to Figure 1-1. But the cutting tool can translate in X and Y directions. The most basic type of turning machine is a lathe. Now, lathes are getting replaced by CNC Turning machines, which are faster and more accurate then the traditional lathes. The CNC Turning machines are controlled by numeric codes. These codes are interpreted by computer attached in the machine and then the computer commands various sections of the machine to do a specific job. The basic operations that can be done on Turning machines are:

* Turning
* Taper turning
* Spherical generation
* Facing
* Grooving
* Parting (in few cases)
* Drilling

Creo Manufacturing 4.0 Black Book

- Boring
- Reaming
- Threading

Milling Machines

Milling machines is a category of machines used for removing material by using a tool perpendicular to the face of the workpiece. In this type of machine, workpiece is held on a bed with the help of fixtures. The tool rotates at a specified speed. This tool can move in X, Y, and Z directions. In some machines, the bed can also translate and rotate like in Mill-turn machines, 5-axis machines, and so on. Milling machines are of two types; horizontal milling machine and vertical milling machine. In Horizontal milling machine, the tool spindle is aligned with the horizontal axis (X-axis). In Vertical milling machine, the tool spindle is aligned with the vertical axis (Y-axis). The Vertical milling machine is generally used for complex cutting processed like contouring, diesinking, and so on. The Horizontal milling machines are used for cutting slots, grooves, gear teeth, and so on. In some Horizontal milling machines, table can moved up-down using the power system. So, using the synchronization of table movement with the rotation of rotary fixture, you can create spiral features. The tools used in both types of milling machines have cutting edge on the sides as well as at the tip.

Drilling Machines

Drilling machines is a category of machines used for creating holes in the workpiece. In Drilling machine, the tool (drill bit) is fixed in a chuck and the chuck can move up-down while rotating at its own axis. The workpiece is fixed on the bed and tool goes down, by motor or by hand, penetrating through the workpiece. There are various types of Drilling machine available like drill presses, cordless drills, pistol grip drills, and so on.

Shaper

Shaper is a category of machines, which is used to cut material in a linear motion. Shaper has a single point cutting tool, which goes back-forth to make linear cut in the workpiece. This type of machine is used to create flat surface. You can create dovetail slots, splines, key slots, and so on by using this machine. In some operation, this machine can be an alternative for EDM.

Planer

Planer is a category of machines similar to Shaper. The only difference is that, in case of Planer machine, the workpiece reciprocates and the tool is fixed.

There are various other special purpose machines (SPMs), which are used for a special type of machining. The machines discussed above are conventional machines. The unconventional machines are discussed next.

Electric Discharge Machine

Electric Discharge Machine is a category of machines used for creating desired shapes on the workpiece with the help of electric discharges. In this type of machines, the tool and the workpiece act as electrodes and a dielectric fluid is passed between them. The workpiece is fixed in the bed and tool can move in X, Y, and Z direction. During the machining process, the tool is brought near to the workpiece. Due to this, a spark is generated between them. This spark causes the material on the tool and workpiece to melt and get separated from the workpiece as well as tool. This separated material is drained with the help of dielectric fluid. There are two types of EDMs which are listed next.

Wire-cut EDM

In this type of EDM, a brass wire is commonly used to cut the material from the workpiece. This wire is held in upper and lower diamond shaped guides. It is constantly fed from a bundle. In this machine, the material is removed by generating sparks between tool and workpiece. A Wire-cut EDM can be used to a plate having thickness upto 300 mm.

Sinker EDM

In this type of EDM, a metal electrode is used to cut the material from the workpiece. The tool and the workpiece are submerged in the dielectric fluid. Power supply is connected to both the tool and the workpiece. When tool is brought near the workpiece, sparks are generated randomly on their surfaces. Such sparks gradually create impression of tool on the workpiece.

Electro Chemical Machine

Electro Chemical Machine is a category of machines used for creating desired shape by using the chemical electrolyte. This machining works on the principles of chemical reactions.

Laser Beam Machine

Laser Beam Machine is a category of machines that uses a beam of highly coherent light. This type of light is called laser. A laser can output a power of up to 100MW in an area of 1 square mm. This type of machines can be used to create accurate holes or shapes on a material like silicon, graphite,diamond, and so on.

These machines were the major machines used in industries. Some of these machines can be controlled by numeric codes. Such type of machines are called NC machines. NC Machines and their working are discussed next.

NC MACHINES

An NC Machine is a manufacturing tool that removes material by following a predefines command set. An NC Machine can be a milling machine or it can be a turning center. NC stands for Numerical Control so, these machines are controlled by numeric codes. These codes are dependent on the controller installed in the machines. There are various controllers available in the market like Fanuc controller, Siemens controller, Heidenhain controller, and so on. The numeric codes change according to the controller used in the machine. These numeric codes are compiled in the form of a program, which is fed in the machine controller via a storage media. The numeric codes are generally in the form of G-codes and M-codes. For understanding purpose, some of the G-codes and M-codes are discussed next with their functions for a Fanuc controller.

Code		Function
G00	-	Rapid movement of tool.
G01	-	Linear movement while creating cut.
G02	-	Clockwise circular cut.
G03	-	Counter-clockwise circular cut.
G20	-	Starts inch mode.
G21	-	Starts mm mode.
G96	-	Provides constant surface speed.
G97	-	Constant RPM.
G98	-	Feed per minute
G99	-	Feed per revolution
M00	-	Program stop
M02	-	End of program
M03	-	Spindle rotation Clockwise.
M04	-	Spindle rotation Counter Clockwise.
M05	-	Spindle stop
M08	-	Coolant on
M09	-	Coolant off
M98	-	Subprogram call
M99	-	Subprogram exit

These codes as well as the other codes will be discussed in the subsequent chapters according to their applications.

As there is a long list of codes which are required to machine a workpiece in the desired size and shape, it becomes very tedious job to create programs manually for each operation. Moreover, it take much time to create a program for small operations on a milling machine. To solve this problem and to reduce the human error, Computer Aided Manufacturing (CAM) was introduced in the market. Various applications of CAM are discussed next.

APPLICATIONS OF COMPUTER AIDED MANUFACTURING

Computer Aided Manufacturing (CAM) is a technology which can be used to enhance the manufacturing process. In this technology, the machines are controlled by a workstation. This workstation can serve more than one machines at a time. Using CAM, you can manage the machining process of more than one machines at one time. Some of the applications of CAM are discussed next.

1. CAM with the combination of CAD can be used to create complex shapes by machining in relatively less time.
2. CAM is used to manage more than one machines at the same time with less human power.
3. CAM is used to automate the manufacturing process.
4. CAM is used to generate NC programs for various types of NC machines.
5. 5-Axis Machining

CAM is generally the next step after CAD (Computer Aided Designing). Sometimes CAE (Computer Aided Engineering) is also required before CAM. There are various software companies that provide the CAM software solutions. PTC is one of the top 10 largest CAM software companies. PTC offers CAM solutions in various small apps. So, depending on your requirements, you can choose the desired apps. These apps can work seamlessly with Creo Parametric or Creo Direct. In this book, Creo Parametric is used as the base application for manufacturing apps.

MANUFACTURING APPS IN CREO PARAMETRIC

There are various apps that can work seamlessly with Creo Parametric. The list of available apps can be found in the **New** dialog box of Creo Parametric. To choose any manufacturing app, start Creo Parametric 4.0 and choose the **New** button available in the **Data** panel of the **Ribbon**; the **New** dialog box will be displayed, refer to Figure-1.

Figure-1. The New dialog box

Select the **Manufacturing** radio button from the **Type** area in the dialog box; the options in the **Sub-type** area will change according to manufacturing category; refer to Figure-2. The options available in Creo Parametric for manufacturing with their short description are given next.

NC Assembly - NC Assembly app is used to create an NC program using different machines like lathe, mill-turn, milling, or EDM.

Expert Machinist - Expert Machinist app is used to create an NC program for a reference model by using milling machine.

CMM - CMM app is used to create a sequence of measurement in reference to the model created in Creo Parametric.

Sheetmetal - Sheetmetal app in manufacturing category is used to create NC sequence for sheetmetal manufacturing. Using Sheetmetal app, you can create various designs in a single manufacturing model.

Cast Cavity - Cast Cavity app is used to create die assemblies and castings.

Mold Cavity - Mold Cavity app is used to simulate the molding process. Using this app, you can create, modify or analyze a mold component.

Harness - Harness app is used to create flat layout of cables on a nail board.

Process Plan - Process Plan app is used to create and publish the documentation of process to create, assemble or manufacture a model.

Figure-2. The New dialog box with the options related to manufacturing

In this book, Expert Machinist and NC Assembly will be used to create NC programs. The method to start Expert Machinist app is discussed next.

STARTING EXPERT MACHINIST

- To start Expert Machinist app, start Creo Parametric, the interface of Creo Parametric will be displayed, as shown in Figure-3.
- Choose the **New** tool from the **Data** panel in the **Ribbon**; the **New** dialog box will be displayed.

Figure-3. Interface of Creo Parametric

- Choose the **Manufacturing** radio button from the **Type** area; the options related to manufacturing will be displayed; refer to Figure-2.
- Choose the **Expert Machinist** from the **Sub-Type** area and enter the desired name of file in the **Name** edit box. By default, the **Use default template** check box is chosen in this dialog box. So, the units and other setting used are default one.
- Clear this check box to choose a different template. If you choose **OK** button after clearing the check box; the **New File Options** dialog box will be displayed, as shown in Figure-4.
- Click on the **OK** button from the dialog box; the **Expert Machinist** interface of Creo Parametric will be displayed, as shown in Figure-5.

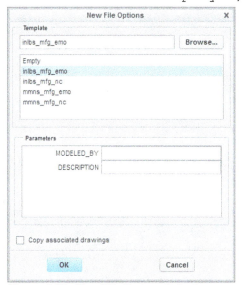

Figure-4. The New File Options dialog box

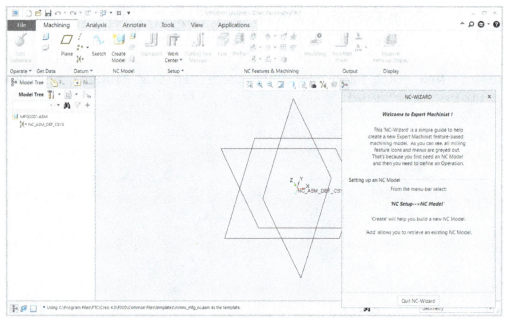

Figure-5. Interface of Expert Machinist

With the Expert Machinist, the **NC-Wizard** dialog box will be displayed; refer to Figure-6. The **NC-Wizard** dialog box provides a step by step guide for creating an Expert Machinist model. If you do not want to display this dialog box, choose the **Quit NC-Wizard** button.

Figure-6. The NC-Wizard dialog box

The other options available in the UI (User Interface) of Creo Expert Machinist are discussed next.

USER INTERFACE OF EXPERT MACHINIST

The user interface of Creo Expert Machinist is displayed, as shown in Figure-7.

RIBBON

Before you start working on Creo Parametric, it is very important for you to understand the default **Ribbon** and tools in the main window. Figure-7 shows various default screen components in Creo Parametric. The **Ribbon** is composed of a series of groups, which are organized into tabs depending on their functionality. The groups in the **Ribbon** that initially appear on the screen are shown in Figure-7. You will notice that all the tools in the groups are not enabled. These buttons will be enabled only after you create a part or open an existing file but the tools that are required for the current session are already enabled. As you proceed to enter one of the modes provided by Creo Parametric, you will notice that the tools required by that mode are enabled. Additionally, to make the designing easy and user-friendly, this software package provides you with a number of groups. Different modes of Creo Parametric display different groups. Some of the frequently used groups are shown in Figure-7.

Figure-7. The user interface of Creo Expert Machinist

TOOLBARS

In Creo Parametric, there are two toolbars, **Quick Access Toolbar** and **In-graphics Toolbar**. The toolbar on the top of the window is called the **Quick Access Toolbar** and the toolbar on the top of the drawing area is called the **In-graphics Toolbar**.

In-graphics Toolbar

The **In-graphics** toolbar has ten buttons by default; refer to shown in Figure-8.

- The first button is **Refit**, which is used to fit a model on the screen.
- The second buttons are **Zoom In** and **Zoom Out**, which are used to enlarge or diminish the model view, respectively.
- The **Repaint** button is used for repainting the screen, which helps in removing any temporary information from the drawing area.
- You can change the display style by using the **Display Style** drop-down.
- Using the **Saved Orientation** button, you can select a view orientation.
- **View Manager** button is used to open the **View Manager** dialog box.

Figure-8. The In-Graphics toolbar

- The next drop-down is **Datum Display Filters**. Using the options in this drop-down, you can toggle the display of datums to on/off.
- You can show or hide the material removal using **Material Removal Display** button.
- The last button in this toolbar is **Spin Center**, which is used to toggle the visibility of the spin center on/off.

File Menu

- The **File** menu contains options shown in Figure-9. These options are used to create a new file, save a file, print the current file, or open an existing file.
- Various other options are also available in the menu to manage the current session and files.
- The **Options** tool is also available in the **File** menu. When you choose this tool, the **Creo Parametric Options** dialog box is displayed. Using this dialog box, you can configure various parameters of Creo Parametric.
- You can also customize the **Ribbon** interface.

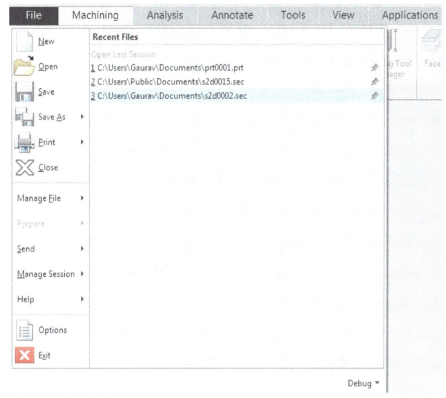

Figure-9. The File Menu

Customizing the Ribbon

- To customize the **Ribbon**, click on the **Options** tool from the **File** menu; the **Creo Parametric Options** dialog box will be displayed.
- Select the **Ribbon** from **Customize** node from the left area of the dialog box; the updated dialog box will be displayed; refer to Figure-10.

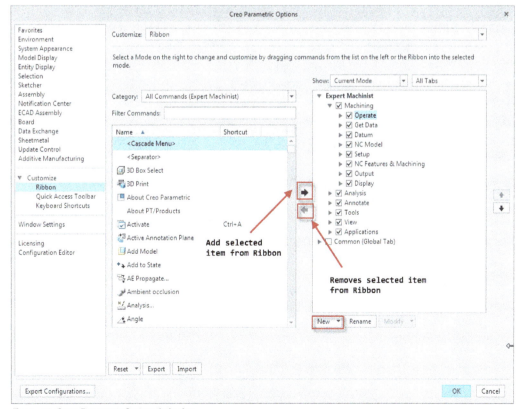

Figure-10. Creao Parametric Options dialog box

- Now, select the specific command from left area of the dialog box. Select the panel from right area of the dialog box, on which you want to add that command and click on **Adds selected item to the Ribbon**.
- Choose the **New Group** or **New Tab** button in **New** drop-down from the button right area of the dialog box to create any new group or tab depending on the requirement; the new group or tab will be added.
- Similarly you can add the command for **Quick Access Toolbar** and buttons for **Keyboard Shortcuts** from **Customize** node.

Search Bar

There are various tools in the help bar which are used to understand the Creo parametric 4.0 software; refer to Figure-11 . The tools of this bar are discussed next.

Figure-11. Help Bar

Minimize the Ribbon

The **Minimize the Ribbon** button is used to minimize the **Ribbon** from the current screen. On again pressing this button, you can maximize the **Ribbon**.

Command Search Button

The **Command Search** button is available at the top right corner of the program window. When you choose this button, the search box is displayed, as shown in Figure-11. You need to enter the first few letters of the command to be searched in the box. When you enter the first letter in the search box, a list of commands starting from that letter will be displayed. You can invoke any command by selecting it from the list, or you can enter the next letter to refine your search.

PTC Learning Connector

This tool is used to learn the tools and commands of Creo Parametric 4.0. To use this tool, click on **PTC Learning Connector** button from Help bar. The **PTC Learning Connector** dialog box will be displayed; refer to Figure-12. You can search the desired topic by using the search box.

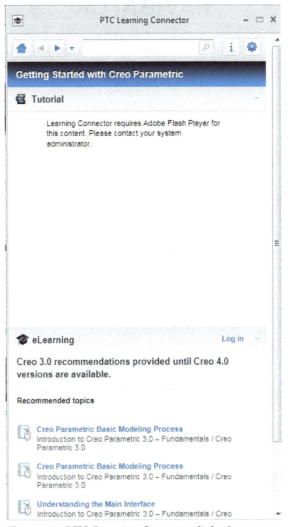

Figure-12. PTC Learning Connector dialog box

Creo Parametric Help

Click on the Creo Parametric Help button from the Help bar. The PTC Creo parametric Help website will be displayed.

Note

*If you want to change the predefined unit system, choose **File > Prepare > Model Properties** from the menu bar to display the **Model Properties** dialog box. In this dialog box, click on the **change** link on the right of the **Units** option under the **Materials** head; the **Units Manager** dialog box will be displayed with the **System of Units** tab chosen. Next, select the desired unit system from the list box and then choose the **Set** button; the **Changing Model Units** message box will be displayed. Choose the **OK** button from the message box; the new unit system will be set and displayed with the red arrow on the left.*

NAVIGATOR

The **Navigator** tool is present on the left of the drawing area and can slide in or out of the drawing area. To make the navigator slide in or out, you need to select the **Toggle the display of navigation area** button on the bottom left corner of the program window. A partial view of the navigator is shown in Figure-13. It has the following functions:

- When you browse files using the navigator, the browser expands and the files in the selected folder are displayed in the browser.
- When you open a model, the **Model Tree** is displayed in the navigation area. The buttons on the top of the navigator are used to display different items in the navigation area.
- The **Model Tree** button is used to display the **Model Tree** in the navigation area. This button is available only when a model is opened.
- The **Folder Browser** button is used to display the folders that are in the local system.
- The **Favorites** button is used to display the contents of the **Personal Favorites** folder.
- Any other location, if available on your system, can also be accessed by using the navigator.

Figure-13. Partial View of the Navigator

Chapter 2

Starting with Expert Machinist

Topics Covered

The major topics covered in this chapter are:

- *Manufacturing Processes in Expert Machinist.*
- *Logical sequence used for creating an NC manufacturing model.*
- *Importing Model for manufacturing.*
- *Creating NC Model for manufacturing.*
- *Retrieving stock from a model*
- *Configuring Units for the NC Model*

MANUFACTURING SEQUENCE IN EXPERT MACHINIST

The Expert Machinist app incorporates various manufacturing processes in a predefined order. The sequence of operations performed in Expert Machinist, to create NC model, is given in the flowchart (Figure-1) given next.

Figure-1. Flow chart for creating manufacturing model

As you can see from the flowchart; the first step for creating a manufacturing sequence is either importing a model of final product or creating model of the final product required after manufacturing. General practice for a CAM engineer is importing the model. The procedure to import a nc model for manufacturing is given next. Note that the procedure to import model created in other software is discussed in Part-II of this book.

IMPORTING NC MODEL

You can import only the NC models in Creo Expert Machinist. The model to be imported must be an assembly file created by any manufacturing app of Creo. The procedure to import the assembly file is given next.

- Click on **Add Model** button from the **NC Model** panel of the **Machining** tab in the **Ribbon** (refer to Figure-2); the **Choose NC Model** dialog box will be displayed, as shown in Figure-3.

Figure-2. The Add Model button

Figure-3. Choose NC Model dialog box

- Browse to the desired file using this dialog box and select the model file to be imported. Now, choose the **OK** button from the dialog box; a new window of Creo Parametric will be displayed along with the **NC Model** Menu Manager, refer to Figure-4.
- Now, using the options in the **Menu Manager**, you can create stock

of the Model to be machined. The procedure to create stock of the model is given next.

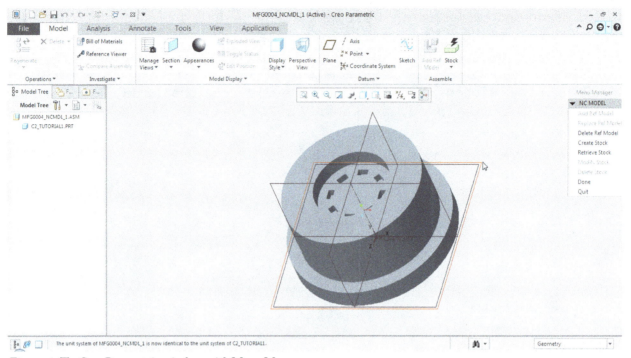

Figure-4. The Creo Parametric window with Menu Manager

Creating Stock

- Click on the **Create Stock** option from the **NC Model Menu Manager**; the Creo Parametric window will be displayed, as shown in Figure-5.
- As you choose the **Create Stock** option, the **Auto Workpiece Creation** contextual tab is added in the **Ribbon**. The options in the **Ribbon** are given next.

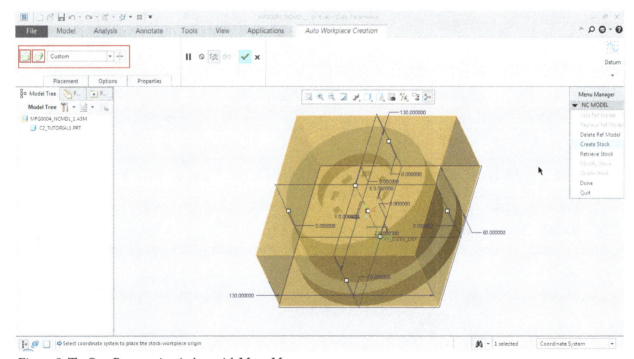

Figure-5. The Creo Parametric window with Menu Manager

Create Rectangular Workpiece

The **Create Rectangular Workpiece** button is available in the left corner of the **Ribbon**.

• On clicking this button, a rectangular jacket is created around the reference model.
• Using the handles displayed on the workpiece, you can add or remove the stock in the specific direction.
• You can also add the material by using the edit box displayed on double-clicking on the value corresponding to the selected handle.

Create Round Workpiece

The **Create Round Workpiece** button is available adjacent to the **Create Rectangular Workpiece** button in the **Ribbon**.

• On clicking this button, a round jacket is created around the reference model, refer to Figure-6.
• You can change the value of stock in different directions using the edit handles in the similar way as done for rectangular stock.

Figure-6. The stock created after choosing the Create Round Workspace option

Stock-workpiece's sub-shape

The **Stock-workpiece's sub-shape** drop-down is available adjacent to the **Create Round Workpiece** button in the **Ribbon** as shown in Figure-5. Options in this drop-down are used to change the shape of stock created around the workpiece. There are two options available in this drop-down, **Custom** and **Envelope**.

• The **Custom** option is used to create a customized workpiece i.e. you can change the size of workpiece by using the edit boxes or edit handles. Figure-7 and Figure-6 show the preview of workpiece created by using the **Custom** option.

*   The **Envelope** option is used to create an envelope around the reference model. Figure 2-7 shows the stock created by the **Envelope** option.

Figure-7. The stock created after choosing the Envelope option

Symmetric about Z axis

The **Symmetric about Z axis** toggle button is available at the right of the **Stock-workpiece's sub-shape** drop-down. If this toggle button is chosen then the stock is created symmetric about the Z axis i.e. you cannot rotate the stock in any direction. If this toggle button is not chosen then you can rotate the stock about any of the three axes (X, Y, and Z).

Options tab

There are various options available in this tab to change settings for stock, refer to Figure-8. Options in this tab are discussed next.

Figure-8. The Options tab

Units

There are two options in the **Units** drop-down; **Inches** and **Millimeters**. The **Inches** option is used to change units of measurement of stock to inch. Whereas the **Millimeters** option is used to change the units of measurement of stock to mm.

Diameter Total

The **Diameter Total** edit box is used to change the value of diameter of stock. This option is available only if you are creating round stock.

Length Total

The **Length Total** edit box is used to change the total length of stock. This edit box is available only when the you are creating round stock.

X Total/Y Total/Z Total

These edit boxes are available only when you are creating rectangular stock.

- The **X Total** edit box is used to specify the thickness of stock in X direction.
- The **Y Total** edit box is used to specify the thickness of stock in Y direction.
- The **Z Total** edit box is used to specify the thickness of stock in Z direction.

Linear Offset

There are two toggle buttons in this area; **Current Offsets** and **Minimum Offsets**.

- On choosing the **Current Offsets** button, you can specify the thickness of stock around the reference model.
- If **Minimum Offsets** toggle button is chosen then you can specify the minimum thickness of stock around the reference model.
- If you are creating rectangular stock, then six edit boxes will be available in the **Linear Offset** area to specify the thickness in X, Y, and Z direction.
- If you are creating round stock, then three edit boxes will be available in the **Linear Offset** area to specify stock in diameter and length.

Rotation Offset

There are three edit boxes available in this area; **About X**, **About Y**, and **About Z**. These edit boxes are used to specify rotation value of stock about X, Y, and Z axes.

- After specifying the desired stock parameters, click on the **Done** button from the **Ribbon** and then choose the **Done** option from the **NC Model** Menu Manager. The assembly environment of Creo Parametric will be displayed; refer to Figure-9 asking you to place the stock as per your machine requirements. Also, you will be prompted to constrain the workpiece by proper constraints.

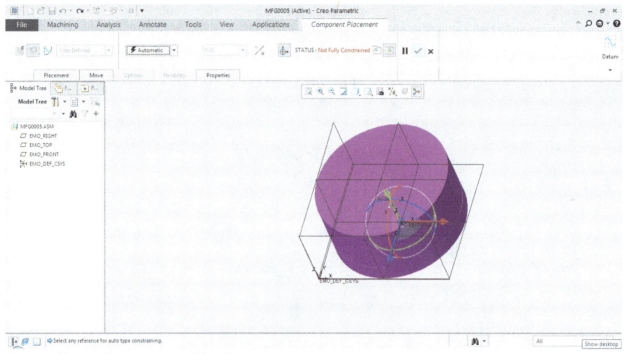

Figure-9. The assembly environment of Creo Parametric

- In our case, we have selected the **Default** option from the **Constrain** drop-down in the **Ribbon**, refer to Figure-10 and choose the **Done** button; the workpiece will be placed at origin.

After assembling the workpiece; the model will be displayed, as shown in Figure-11.

Figure-10. The Constrain drop-down

CREATING NC MODEL

Earlier, we have added the NC model already created in Creo. Now, we will learn how to create NC model in Expert Machinist. The procedure to create NC model is given next.

- To create an NC Model, choose the **Create Model** button from the **NC Model** panel of the **Machining** tab in the **Ribbon**. You will be prompted to specify the name of the model to be created.
- Enter the name of the nc model. The **Open** dialog box will be displayed; refer to Figure-12.
- Select an assembly or a part to be added as reference part for workpiece and choose the **Open** button. If the selected entity has more than one instances then the **Select Instances** dialog box will be displayed, as shown in Figure-13.
- Select the desired instance of the model. The reference model will be displayed in the modeling area and **NC Model** Menu Manager will be displayed. You can create the workpiece by using the Menu Manager, as discussed earlier.

Figure-11. The model after assembeling the workpiece

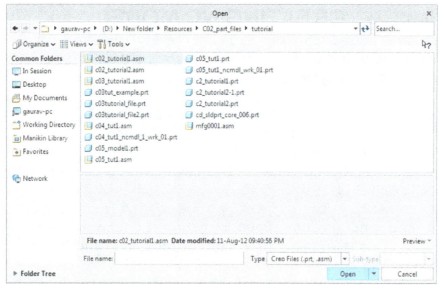

Figure-12. The Open dialog box

MODIFY MODEL

You can modify the NC Model present in the drawing area as per your requirement by using the **Modify Model** tool. The procedure is given next.

- To modify the model, choose the **Modify Model** tool from the **NC Model** panel of the **Machining** tab in the **Ribbon**. You will be prompted to choose an NC Model and **NC Model** Menu Manager will be displayed.
- Choose the **Modify Stock** option from the Menu Manager. The **Auto Workpiece Creation** contextual tab will be displayed. Options in this tab have already been discussed.

Figure-13. The Select Instances dialog box

DELETE MODEL

You can delete one or more of the NC Model present in the drawing area by using the **Delete Model** tool. The procedure is given next.

- To delete an NC Model, choose the **Delete Model** tool from the **NC Model** panel of the **Machining** tab in the **Ribbon**. You will be prompted to select an NC Model from the drawing area.
- If you delete the parent model then the **Child** Menu Manager will also be displayed, as shown in Figure-14. Various options are available in the Menu Manager like to delete, suspend or freeze the NC Model. Select the desired option to delete one or all the child features.

RETRIEVING STOCK

You can add an assembly or a model as a stock by using the **Retrieve Stock** option while creating NC model. The procedure to use this tool is discussed next.

*   Click on the **Retrieve Stock** option from the **Menu Manager** displayed while creating a workpiece, refer to Figure-15. The **Open** dialog box will be displayed, refer to Figure-16 and you will be prompted to selected a file for creating stock.

Figure-15. The NC MODEL Menu Manager

Figure-14. The Child Menu Manager

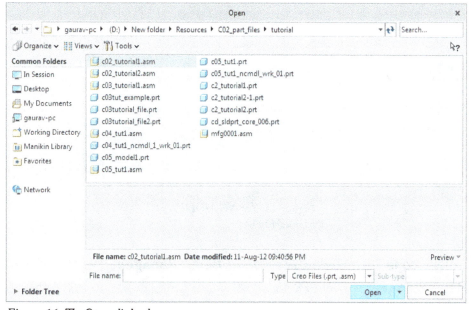

Figure-16. The Open dialog box

*   Select the desired stock file and click on **Open** button from the dialog box; the **Component Placement** tab will be activated and you will be prompted to assemble the stock with the reference part, refer to Figure-17.

Figure-17. The application window with the Component Placement tab activated

- Assemble the stock with the reference part using proper constraints and then choose the **Done** button ✔. The Menu Manager will be displayed, as shown in Figure-18. (Note: Assembly constraints are out of scope of this book. To know more about Assembly constraints, please refer to another book of author- **Creo Parametric 4.0 Black Book**).

Figure-18. The NC MODEL Menu Manager

- Choose the **Done** button from the **Menu Manager**; again the **Component Placement** tab will be displayed and you will be prompted to assemble the workpiece with the NC Coordinate system.
- Select the **Default** constrain from the **Constraint** drop-down list and choose the **Done** button; the workpiece will be creating in the modeling area along with the reference part.

Setting Units for the NC Model

While working in the **Expert Machinist** environment, you can change the units of model any time. The procedure to use this tool is discussed next.

- Click on the **Units** tool from the expanded **Setup** panel in the **Machining** tab; the **Units Manager** dialog box will be displayed, as shown in Figure-19.
- There are two tabs available in this dialog box, **Systems of Units** and **Units**. Options available in these tabs are discussed next.

Figure-19. The Units Manager dialog box

System of Units

Options in this tab are displayed as shown in Figure-19. These options are discussed next.

Set

A blue arrow is displayed adjacent to the unit system selected. In our case **Inch lbm Second** is selected. So, a blue arrow is displayed adjacent to the unit system.

- To set any other unit system, select the unit system and click on the **Set** option from the right in the dialog box. The **Changing Model Units** dialog box will be displayed; refer to Figure-20.

Figure-20. Changing Model Units dialog box

- Select the required option from **Changing Model Units** dialog box. If by mistake you have designed your model in inches or mm and you want the system to interpret 1″ as 1 mm or 1 mm as 1″, then click on the **Interpret dimensions** radio button. If you want to convert your model unit system from inch to mm or vice-versa, then click on the **Convert dimensions** radio button.

- Click on **OK** button. The red arrow will shift to that selected unit system and the unit system will be set.

New

The **New** option is used to create a user defined unit system. The procedure to use this tool is discussed next.

- Click on **New** button from **Units Manager** dialog box. The **System of Units Definition** dialog box will be displayed as shown in Figure-21.

Figure-21. The System of Units Definition dialog box

- Specify the desired name in the **Name** edit box.
- Select the radio button adjacent to the desired type of unit system from the **Type** area. The options available in the **Units** area will modify accordingly. Select the desired option from the **Length, Mass/ Force, Time,** and **Temperature** drop-downs.
- Click on the **Info** button to check the derived units. The **INFORMATION WINDOW** is displayed as shown in Figure-22. All the derived units from the current unit system are displayed in this window.
- After checking the derived units, choose the **Close** button from the window and choose the **OK** button from the **System of Units Definition** dialog box.; The new unit system will be displayed in the unit system list in the **Systems of Units** tab.

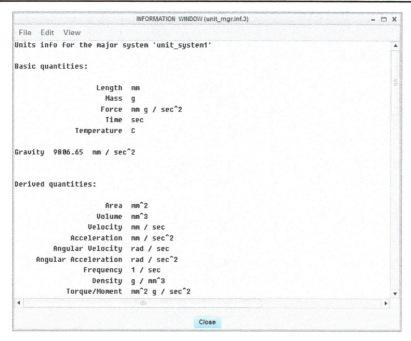

Figure-22. The INFORMATION WINDOW

Copy

The **Copy** option is used to create a copy of the already created unit system. The procedure is given next.

- Click on the **Copy** button, the **Copy System of Units** dialog box is displayed, as shown in Figure-23.
- You can specify the desired name in the **To new unit** edit box available in the **Copy System of Units** dialog box.
- After specifying the parameters, click on the **OK** button to create a copy of the currently set unit system.

Figure-23. The Copy System of Units dialog box

Edit

The **Edit** option is used to edit an already created unit system apart from the default unit systems. The procedure is given next.

- Click on the **Edit** button, the **System of Units Definition** dialog box is displayed. This dialog box is similar to the one displayed on choosing the **New** option.

Delete

The **Delete** option is used to delete an already created unit system apart from the default unit systems. By using this option, you can delete only the custom created unit system. To delete a custom unit system, choose the **Delete** option from the dialog box. The **Unit Manager - Confirmation** dialog box will be displayed. Click on the **Confirm** button. The unit will be displayed.

Info

The **Info** option is used to display information about the selected unit system.

- Click on the **Info** button, the **INFORMATION WINDOW** is displayed. Options in this window have already been discussed.

TUTORIAL 1

In this tutorial, you will import the reference model as shown in Figure-24 and you will create a cylindrical workpiece having thickness of 2 mm over the reference model. The model file for reference model is available in the Resources of this book.

Figure-24. The reference model to be imported

Steps required to complete the tutorial:

1. Copy the model from the resource kit in your local drive.
2. Open the **Expert Machinist** application window by using the metric template.
3. Create a new workpiece using the reference model copied from the resource kit.

Copying the reference model

1. Extract the resources kit folder provided with the book at the desired location.
2. Browse to the location **\resources\Chapter2\data** and copy the file in the **Creo Manufacturing\Chapter2\tutorial** folder in your local drive.
3. Set the Creo Manufacturing folder as a working directory.

Tip

To set a folder as working directory, select **Folder Browser** option from the **Navigator** area and expand the **Folder Tree** node. Select the desired folder from the **Folder Tree** and right click on it; a shortcut menu is displayed. Choose the **Set Working Directory** option from the shortcut menu; the folder will become the current working directory.

Starting Expert Machinist

1. Open the Creo Parametric application by double clicking on the Creo Parametric icon on the desktop.
2. Choose the **New** button from the **Data** panel in the **Home** tab of the **Ribbon**; the **New** dialog box will be displayed as shown in Figure-25.

Figure-25. The New dialog box

3. Choose the **Manufacturing** radio button from the **Type** area and **Expert Machinist** from the **Sub-type** area.
4. Specify the name as **c02 tutorial1** in the **Name** edit box and clear the **Use default template** check box from the dialog box.
5. Choose the **OK** button from the dialog box. The **New File Options** dialog box is displayed as shown in Figure-26.

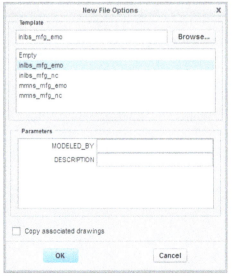

Figure-26. The New File Options dialog box

6. Select the **mmns mfg nc** option from the list of templates displayed and choose the **OK** button from the dialog box; the **Expert Machinist** environment of the Creo Parametric will be displayed with metric units selected.

Creating a new NC Model

- Choose the **Create Model** tool from the **NC Model** panel of **Machining** tab in the **Ribbon**. A message box will be displayed as shown in Figure-27 and you will be prompted to enter the name of NC model.

Figure-27. The message box displayed

- Enter the name as **nc tutorial1** in the message box and choose the ✔ button from the message box. The **Open** dialog box will be displayed; refer to Figure-28.

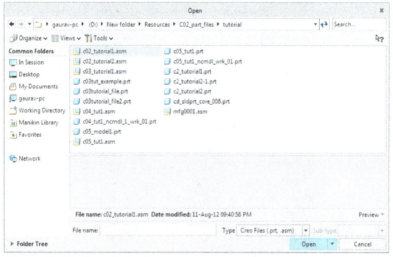

Figure-28. The Open dialog box

- Browse to the location **Creo Manufacturing\Chapter2\tutorial**, select the c2 tutorial1.prt file and choose the **Open** button from the dialog box. The application window will be displayed as shown in Figure-29.

Figure-29. The application window after selecting the file

- Click on **Create Stock** option from the **NC MODEL** Menu Manager. A rectangular stock is created around the reference model as shown in Figure-30.

Figure-30. The rectangular block created around the reference model

- Click on the **Create round workpiece** button from the **Auto Workpiece Creation** contextual tab. The model will be displayed as shown in Figure-31.

Figure-31. The model after choosing the Create Round Workpiece button

- Click on the **Options** tab. A flyout is displayed.
- Enter **90** in the **About X** edit box available in the **Rotation Offsets** area of the flyout. The model will be displayed as shown in Figure-32.

Figure-32. The model after specifying 90 in the About X edit box

- Specify the value of diameter and length as **132** and **62** in the **Diameter Total** and **Length Total** edit boxes, respectively. Make sure that millimeters unit is selected in the **Units** drop-down list.
- Click on the **Done** button ✔ from the **Ribbon** to create the workpiece. The workpiece will be displayed as a green transparent round block; refer to Figure-33.

Figure-33. The workpiece with the reference model

- Click on the **Done** option from the **NC MODEL** menu manager; the Expert Machinist environment will be displayed along with the **Component Placement contextual** tab selected. You will be prompted to place the workpiece with reference part in Machine coordinate system.
- Click on the **Default** option from the **Component Placement** drop-down list in the **Ribbon** and click on the **Done** button.

TUTORIAL 2

In this tutorial, you will import a reference model provided in the resources as c02 tutorial2 and you will create a workpiece by using a retrieved model provided in resources as c2 tutorial2-1.prt. The model file for reference model and retrieving workpiece model are available in the Resources kit.

Steps required to complete the tutorial:
* Copy the model from the resource kit in your local drive.
* Open the Expert Machinist application window by using the metric template.
* Retrieving the model provided for creating workpiece.

Copying the reference model

* Extract the resources kit folder provided with the book at the desired location.
* Browse to the location **\resources\chapter2\data** and copy the file in the **Creo Manufacturing 4.0\Chapter2\tutorial** folder in your local drive.
* Set Creo Manufacturing 4.0 folder as a working directory.

 Tip

To set a folder as working directory, select **Folder Browser** option from the **Navigator** area and expand the **Folder Tree** node. Select the desired folder from the **Folder Tree** and right-click on it; a shortcut menu will be displayed. Choose the **Set Working Directory** option from the shortcut menu; the folder will become the current working directory.

Starting Expert Machinist

* Open the Creo Parametric application by double clicking on the Creo Parametric icon on the desktop.
* Click on the **New** button from the **Data** panel in the **Home** tab of the **Ribbon**. The **New** dialog box will be displayed, as shown in Figure-34.

Figure-34. The New dialog box

- Click on the **Manufacturing** radio button from the **Type** area and **Expert Machinist** from the **Sub-type** area.
- Specify the name as **c02 tutorial2** in the **Name** edit box and clear the **Use default template** check box from the dialog box.
- Click on the **OK** button from the dialog box. The **New File Options** dialog box is displayed as shown in Figure-35.

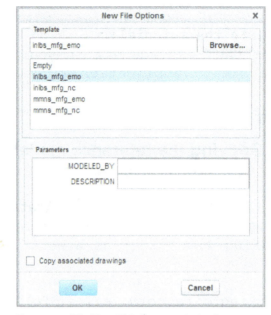

Figure-35. The New File Options dialog box

- Select the **mmns mfg nc** option from the list of templates displayed and choose the **OK** button from the dialog box. The Expert Machinist environment of the Creo Parametric will be displayed with metric units selected.

Creating a new NC Model

- Click on the **Create Model** tool from the **NC Model** panel of **Machining** tab in the **Ribbon**. A message box will be displayed, as shown in Figure-36 and you will be prompted to enter the name of NC model.

Figure-36. The message box displayed

- Enter the name as **nc tutorial2** in the message box and choose the ✔ button from the message box; the **Open** dialog box will be displayed along with **NC MODEL** menu manager, refer to Figure-37.
- Browse to the location **Creo Manufacturing\Chapter 2\tutorial**, select the c2 tutorial2.prt file and choose the **Open** button from the dialog box. The application window will be displayed, as shown in Figure-38.

Figure-37. The Open dialog box along with NC MODEL menu manager

Figure-38. The application window after selecting the file

- Click on the **Retrieve Stock** option from the **NC MODEL** menu manager. The **Open** dialog box will be displayed.
- Browse to the location **Creo Manufacturing 4.0\Chapter 2\tutorial** and select the c02 tutorial2-1.prt file.
- Click on the **Open** button from the dialog box. The application window will be displayed, as shown in Figure-39.

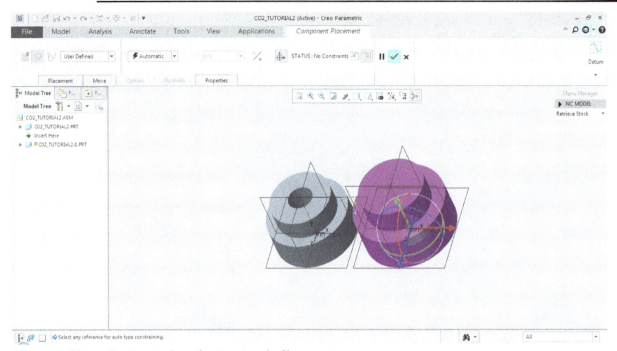

Figure-39. The application window after inserting the file

- Apply coincident constraint on all the three planes of retrieved model with the reference model and choose the **Done** button from the **Ribbon**. The retrieved model for workpiece with reference model will be displayed as shown in Figure-40.

Figure-40. The model after retrieving the model of workpiece

- Choose the **Done** option from the **NC MODEL** menu manager. You will be prompted to place the nc model in the machining coordinate system.
- Choose the **Default** option from the **Constrain** drop-down list and choose the **Done** button from the **Ribbon**; the model will be displayed as shown in Figure-41.

Figure-41. The refrence model with workpiece

PRACTICE 1

Create a workpiece for the model given in Figure-42. Dimensions of the model are given in Figure-43. You can also find this model in the resources kit.

Figure-42. Practice 1

Figure-43. Dimension for Practice 1

PRACTICE 2

Create a workpiece for the model given in Figure-44. Dimensions of the model are given in Figure-45. You can also find this model in the exercise area of the resources kit.

Figure-44. Practice2

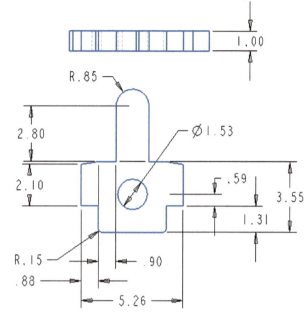

Figure-45. Dimension for Practice 2

FOR STUDENT NOTES

Chapter 3

Tool Setting

Topics Covered

The major topics covered in this chapter are:

- *Basics of Tooling.*
- *Various tools available for milling and parameters.*
- *Creating a tool setup.*
- *Saving a tool setup*
- *Importing a tool setup.*
- *Creating a user defined tool*
- *Checking for minimum radius*
- *Checking for shortest edge*

TOOLING OF A MACHINE

Tool is the main cutting member of machine. All the output of a machine depends on the type of tool selected. The tool selection in return depends on two main factors: Machine to be used and type of machining. Like, for lathe machines generally HSS bar is used, for CNC turning machine tungsten coated bits are used. Figure 3-1 shows the components of a tool required in CNC Milling Machine.

TOOLS AVAILABLE FOR MACHINING IN EXPERT MACHINIST

Expert Machinist is mainly used for creating programs for milling machines. Various tools that are generally used on a milling machine are discussed next.

End Mill

End Mill is a type of milling cutter used for removing material in axial as well as other directions. A few end mill can not cut material in axial direction which means they can not be used for drilling purpose. In a milling machine, End Mill cutter is used for operations like profiling, facing, plunging, and tracing. Parameters required to defined an end mill cutter in Expert Machinist are displayed in Figure-1. Along with the dimensions shown in Figure-1, you need to specify number of flutes, dimension units and the material of tool in Expert Machinist.

Figure-1. An End Mill with parameters

Ball Mill

Ball Mill, also called Ball Nose mill, is a type of milling cutter having hemispherical at the bottom. These type of cutters are used to create 3D contour shapes in workpiece. These type of cutters are used in machining moulds and dies having 3D curves. Figure-2 shows the shape of a Ball Mill cutter and parameters required in Expert Machinist.

Figure-2. A Ball Mill with parameters

Bull Mill

Bull Mill, also called Bull Nose mill, is a type of milling cutter having fillet of specified value at the corner edges. This type of cutter is used for creating filleted corners in the workpiece. Generally, dies and moulds having internal fillets are machined by using this type of cutters. Figure-3 shows a Bull mill with the parameters required in Expert Machinist.

Figure-3. A Bull Mill with parameters

Taper Mill

Taper Mill is a type of milling cutter having radius at corner edges and taper at the cutting length. It is similar to the Bull Mill having the only difference that it has taper at the cutting length. This type of cutter is used for creating profiles having taper at the walls as well as fillet at the bottom. Figure-4 shows a Taper mill with the parameters required in Expert Machinist.

Figure-4. A Taper Mill with parameters

Taper Ball Mill

Taper Ball Mill is a type of milling cutter having curvature of a ball at the end. It is similar to a ball mill with the only difference that it has tapers cutting edges. This type of cutter is used for creating profiles having taper at the walls as well as fillet at the bottom. Figure-5 shows a Taper mill with the parameters required in Expert Machinist.

Figure-5. A Taper Ball Mill with parameters

Drill

Drill is a type of milling tool used to create holes in a workpiece. A drill can not be used for creating profile on the workpiece. Moreover, a drill can cut along the perpendicular axis because the bottom edges are the cutting edges. You should not use the side edges of a drill for cutting purpose. Figure-6 shows a drill with the parameters required in Expert Machinist.

Figure-6. A Drill with parameters

Basic Drill

A basic drill is similar to the drill discussed earlier. The only difference is that it does not have tip. So, you need not to specify the tip length and tip diameter.

Countersink

A countersink is a type of drill used to create counter sunk holes in the workpiece. It is similar to the drill discussed earlier. The only difference is that its tip angle is 90 degree by default.

Spot Drill

A Spot drill is similar to basic drill discussed earlier. The only difference is that its tip angle is 90 degree by default.

Tap

Tap is a type of milling cutter used to create threads in a hole. Figure-7 shows a Tap with the parameters required in Expert Machinist.

Figure-7. A Tap with parameters

Reamer

Reamer is a type of milling cutter used to increase the diameter of a hole. The total increase in diameter is less than 5% of the diameter of the hole. Also, the tolerance of the hole diameter created by reaming is less than 0.1 mm. So, a reamer is used to create accurate holes. Figure-8 shows a Reamer with the parameters required in Expert Machinist.

Figure-8. A Reamer with parameters

Boring Bar

Boring Bar is a type of milling cutter used to increase the diameter of a hole. Figure-9 shows a Boring Bar with the parameters required in Expert Machinist.

Figure-9. A Boring Bar with parameters

Center Drill

Center drill is a type of milling cutter used to make center holes. Mainly it is used to provide starting hole for a larger diameter hole.

Back Spotting

Spot drill is used to provide a starting point for a common drill. The process of drilling such a point is called back spotting. Figure-10 shows a Spot drill with the parameters required in Expert Machinist.

Figure-10. A Spot drill with parameters

Side Milling

Side Milling tool is a type of milling cutter used to remove material of large area from face and walls of the workpiece. Figure-11 shows a Side Milling tool with the parameters required in Expert Machinist.

Figure-11. A Side Milling tool with parameters

Key Cutter

Key Cutter is a type of milling cutter used to cut slots for key in a workpiece. Shape of this tool is similar to the Side Milling tool but the Cutting length of this tool is less and diameter is more.

Lollipop

Lollipop is a type of milling cutter used to create undercut in a workpiece. This tool can also be used for machining curvatures on a workpiece.

Grooving

Grooving tool is a type of milling cutter used to create undercut in a workpiece. This tool is also used to create deep grooves in the workpiece.

Corner Rounding

As the name suggests, this type of milling tools is used to create rounds at the corners of the workpiece. Shape of the tool is displayed as shown in Figure-12.

Figure-12. A Corner Rounding tool

Chamfering

As the name suggests, this type of milling tools is used to create chamfers at the edges of the workpiece.

Thread Mill

Thread Mill is a type of milling tools is used to create threads in a hole created earlier in the workpiece.

Plunge Mill

Plunge Mill is a type of milling tools is used for heavy plunging operations. The Material Removal Rate of this tool generally high.

Multi Tip

This is a user specified tool created in Expert Machinist. This tool has a geometry specified by the user. Also, it has cutting edges specified by the user.

CREATING A TOOL SETUP

A tool setup is the arrangement of tools required for an operation on a specific machine. So, before creating a tool setup, you must have an operation and a machine already created in Creo Expert Machinist. The procedure of creating a tool setup is discussed next.

• To create a machine, click on the **Machine Tool Manager** tool from the **Work Center** drop-down in the **Setup** group of the **Machining** tab of the **Ribbon**. The **Milling Work Center** dialog box will be displayed; refer to Figure-13.

Figure-13. The Milling Work Center dialog box

- Click on the **OK** button from the dialog box to create the machine.
- Click on the **Operation** tool from the **Setup** group of the **Machining** tab of the **Ribbon**. The **Operation** contextual tab will be displayed in the **Ribbon**; refer to Figure-14.

Figure-14. Partial view of the Operation contextual tab

- Click on the **No Work center** drop-down displayed in the dashboard and select the **MILL01** from the list, you will be prompted to select a coordinate system.
- Select the default coordinate system and choose the **Done** button from the dashboard.
- Now, the **Cutting Tool Manager** tool will become activate in the **Ribbon**. Click on the **Cutting Tool Manager** tool from the **Setup** group in the **Machining** tab of the **Ribbon**; the **Tools Setup** dialog box will be displayed, as shown in Figure-15.

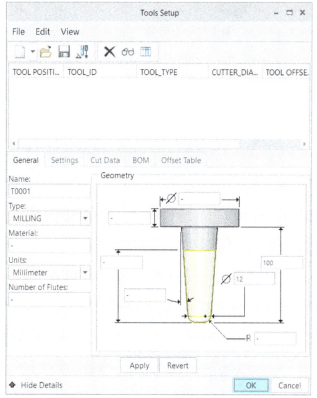

Figure-15. The Tools Setup dialog box

There are five tabs available in this dialog box; **General**, **Setting**, **Cut Data**, **BOM**, and **Offset Table**. Options available in these tabs are discussed next.

General Tab

The options in the **General** tab are used to specify general information about the tool. The options available in this tab are discussed next.

Name

The **Name** edit box is used to specify the name of the tool.

Type

Options in the **Type** drop-down list are used to specify the type of tool. Tools in this drop-down list are available according to the machine selected for the Work center. The type of tools available in this drop-down list have already been discussed.

Units

The options in the **Units** drop-down list are used to specify the unit of measurement for the tool. There are five options available in this drop-down list; **Inch**, **Foot**, **Millimeter**, **Centimeter**, and **Meter**.

Number of Flutes

The **Number of Flutes** edit box is used to specify the number of flutes of the tool

After specifying the above settings choose the **Apply** button to create the tool. On choosing the **Apply** button, the tool will be added to the tool list displayed above in the dialog box.

Settings Tab

The options in the **Settings** tab are used to specify settings for tool number, offset number, gauge length, and so on; refer to Figure-16. Various options available in this tab are discussed next.

Figure-16. The Settings dialog box

Tool Number

The **Tool Number** edit box is used to specify the number of the tool. This tool number is used by the machine while creating NC program to identify each tool.

Offset Number

The **Offset Number** edit box is used to specify the offset number for the current tool. If you have earlier worked on the NC machine then it would be easy for you to understand this number.

 Tip

Offset number is the number specified in the Offset table of the tool compensation in the NC machine. Offset table consists of offset values for the respective offset number. Offset is a value specified in the machine for wear compensation. When tool starts machining then due to friction and heat, the edge of tool starts wearing hence it cuts lesser material than it was assumed to cut. So, to compensate this tool wear this offset value is specified.

Gauge X Length

The **Gauge X Length** edit box is used to specify the gauge length of the tool tip along the X axis.

Gauge Z Length

The **Gauge Z Length** edit box is used to specify the gauge length of the tool tip along the Z axis.

Comp. Oversize

The **Comp. Oversize** edit box is used to specify difference between the actual maximum tool diameter and the cutting diameter specified in the programming. This value is used for gouge checking.

Comments

The **Comments** edit box is used to write comments about the tool and is using techniques.

Cut Data Tab

The options in **Cut Data** tab are used to specify the cutting data for the tool. On choosing this tab, the options are displayed as shown in Figure-17. The options in this tab are discussed next.

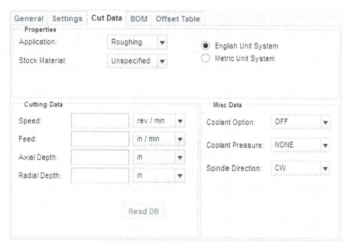

Figure-17. The Cut Data tab

Application

The **Application** drop-down is used to specify the purpose for which the tool is to be used. There are two options available in this drop-down i.e. Roughing and Finishing.

Stock Material

The **Stock Material** option is used to specify the material to be used for the stock.

• Click on **Stock Material** drop-down, a list of material available is displayed. This list depends on the parameters specified in the mfg wp material list.xml file in the program directory.
• Select the desired material from the list.

English Unit System / Metric Unit System

These radio buttons are used to set the unit system in which the parameters of the tool will be specified.

• Select **English Unit System** radio button, if you want to specify the required parameters for the tool in Inches.

• Select the **Metric Unit System** radio button, If you want to specify the required parameters for the tool in Millimeters.

Speed

The **Speed** option is available in the **Cutting Data** area of the **Tools Setup** dialog box. This option is used to specify the rotational speed at which the tool will perform cutting. You can specify the rotational value in two ways, revolution per minute or meters/feet per minute.

Feed

The **Feed** option is available in the **Cutting Data** area of the **Tools Setup** dialog box. This option is used to specify the feed rate at which the tool will cut through the workpiece. You can specify value for feed rate in two ways, mm/inch per minute or mm/inch per tooth.

Axial Depth

The **Axial Depth** option is available in the **Cutting Data** area of the **Tools Setup** dialog box. This option is used to specify the depth along axis of tool in the workpiece at which the cutting operation will be performed.

Radial Depth

The **Radial Depth** option is available in the **Cutting Data** area of the **Tools Setup** dialog box. This option is used to specify the depth along the radial direction of tool in the workpiece at which the cutting operation will be performed.

Coolant Option

The **Coolant Option** drop-down is available in the **Misc Data** area of the **Tools Setup** dialog box. The options in the **Coolant Option** drop-down are used to specify the condition of coolant during machining process. There are six options available in this drop-down; **Flood**, **Mist**, **OFF**, **ON**, **TAP**, and **THRU**.

• If you choose the **Flood** option, the coolant will start flowing through all the passages available in the machine for along the tool.
• **Mist** option is used if you require a mist of coolant at the workpiece.
• **OFF** option is used when you do not require coolant while cutting material.
• **ON** option is used to make the flow of coolant at standard settings.
• **TAP** option is used to make the flow of coolant like a tap, i.e. natural flow.
• **THRU** option is used to make the flow coolant through the tool holder as well as through the vents available for the workpiece.

Coolant Pressure

The **Coolant Pressure** drop-down is available in the **Misc Data** area of the **Tools Setup** dialog box. The options in this drop-down list are used to specify the pressure at which the coolant will flow while cutting operation is performed. There are four options available in this drop-down; **NONE**, **LOW**, **MEDIUM**, and **HIGH**.

Spindle Direction

The **Spindle Direction** drop-down is also available in the **Misc Data** area of the **Tools Setup** dialog box. The options in this drop-down list are used to specify the direction in which the tool spindle will rotate while cutting.

BOM Tab

The options in **BOM** tab are used to specify the bill of material for the tool. A bill of material for tools provide the information like tool type, quantity available, and so on.

Offset Table Tab

The options in the **Offset Table** tab are used to set a table of offset values respective to the offset numbers; refer to Figure-18.

- This offset number is used by the tool for specifying the tool wear compensation. You can add as much offset numbers as required.
- To add more than one offset numbers, select the **Enable Multiple Tips** check box. The plus (**+**) button adjacent to the check box will get activated.
- Click on the plus button to add more offset numbers in the table.

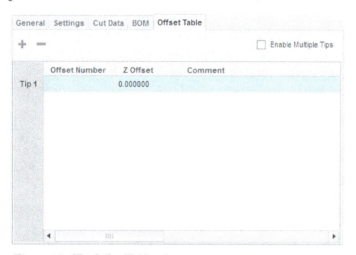

Figure-18. The Offset Table tab

Save tool parameters to a file

This tool is used to save the parameters specified for the selected tool in the form of an xml file. The procedure to save tool parameters to a file is discussed next.

- Click on the **Save tool parameters to a file tool** from the toolbar available at the top of the **Tools Setup** dialog box.
- The tool will be saved in the form of an xml file in the default directory. The tool will be saved with the file name specified in the **Name** edit box of the **General** tab in the dialog box.

New drop-down

The tools in the **New** drop-down are used to create new tool setup based on the templates available in this drop-down. Figure-19 shows the tools available in this drop-down. Choose any of the tool available in this drop-down; the template will be loaded on the basis of the selected tool.

Figure-19. The New drop-down

Retrieve tool from disk

The **Retrieve tool from disk** tool is used to add a tool in the system from a file. The description about the tool is saved in the form of an xml file. The procedure to retrieve the tool from disk is given next.

- Click on the **Retrieve tool from disk** tool from the tools available below the menu bar in the dialog box. The **Open** dialog box will be displayed as shown in Figure-20.
- Browse to the desired file and then click on **Open** button from the dialog box. The tool description in the selected file will be added in the tool list.

Figure-20. The Open dialog box

Delete a tool table entry

The **Delete a tool table entry** tool is used to delete an entry from the table.

Display tool in a separate window
based on current data settings

The **Display tool in a separate window based on current data settings** tool is used to display the selected tool in a separate window. On choosing this tool, a separate window is displayed refer to Figure-21.

Figure-21. The tool window

The tool in this window is displayed with the parameters specified in the edit boxes in **Geometry** area of the **General** tab. Click on the **Close** button from the dialog box to exit.

Customize Tool Parameters Columns

The **Customize Tool Parameters Columns** tool is used to customize the columns in the tool table of the **Tool Setup** dialog box.

• Click on the **Customize Tool Parameters Columns** tool. The **Column Setup Builder** dialog box will be displayed as shown in Figure-22.

Figure-22. The Column Setup Builder dialog box

There are two areas in the **Column Setup Builder** dialog box i.e. **Not Displayed** and **Displayed**. Select the desired option from list in the **Not Displayed** area and then choose the **>>** button available in the middle of both areas. The selected option will be added in the columns displayed in **Tools Setup** dialog box. You can add as many number of columns as are available in the list.

CHECKING FOR MINIMUM RADIUS

Whenever a tool setup is created for the workpiece, it is required to check for minimum radius in the workpiece that is to be created by the tool. On the basis of this minimum radius, the corner radius of the tool is decided. The procedure to check for minimum radius is discussed next.

- Click on the **Radius** tool from the expanded **Inspect Geometry** group in the **Analysis** tab of the **Ribbon**, the **Radius** dialog box will be displayed as shown in Figure-23. Also, you will be prompted to select a geometry for inspection.

Figure-23. The Radius dialog box

- Now, select the **Component** option from the **Filter** drop-down available at the bottom of the application window; refer to Figure-24.
- Select the component from the **Model Tree** (in the left of the application window) for which you want to check the minimum radius.
- Figure-25 shows a model with its minimum radius displayed both from inside and outside. Figure-26 shows the respective dialog box.

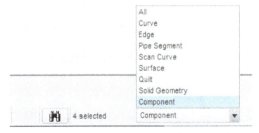

Figure-24. The Component option in the Filter drop-down

Figure-25. The model with minimum radius curve selected

Figure-26. The Radius dialog box

- You can display the Inside radius only or the Outside radius only by using the respective option from the **Plot** drop-down in the **Radius** dialog box. Minimum radius of the curve is highlighted in the dialog box in an ellipse.

CHECKING FOR MAXIMUM DEPTH OF CUT IN THE WORKPIECE

Whenever a tool setup is created for the workpiece, it is also required to check for the maximum depth of cut that is to be achieved. On the basis of this depth, we decide the total length of the tool and cutting length of the tool to avoid any accident in the machine.

- To check for the maximum depth of cut, click on the **Distance** tool from the **Measure** drop-down in the **Measure** panel of the **Analysis** tab in the **Ribbon**, hold the **CTRL** key and select the edge of bottom face and the corresponding edge of top face of the cavity that can cause accident; the distance will be displayed in a box in the Modeling area.

TUTORIAL 1

In this tutorial, you will import a reference model as shown in Figure-27 and you will import the workpiece reference model. Then you will create a tool setup for various operations on the workpiece. The model file for reference model is available in the Resources kit.

Figure-27. Reference model for tutorial 1

Steps required to complete the tutorial
* Copy the model from the resource kit in your local drive.
* Open the Expert Machinist application window by using the metric template.
* Create a new workpiece using the reference model provided.
* Find out various operations to be performed on the workpiece.
* Create a tool setup for various operations

Copying the reference model
* Extract the resources kit folder provided with the book at the desired location.
* Browse to the location **\resources\chapter3\data** and copy the files in the **Creo Manufacturing\Chapter3\tutorial** folder in your local drive.
* Set the Creo Manufacturing folder as a working directory.

Starting Expert Machinist
* Open the Creo Parametric application by double-clicking on the Creo Parametric icon on the desktop.
* Choose the **New** button from the **Data** panel in the **Home** tab of the **Ribbon**. The **New** dialog box will be displayed as shown in Figure-28.

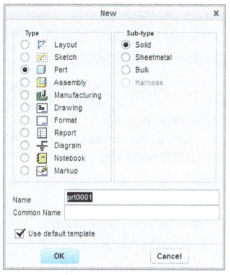

Figure-28. The New dialog box

- Choose the **Manufacturing** radio button from the **Type** area and **Expert Machinist** from the **Sub-type** area.
- Specify the name as **c03 tutorial1** in the **Name** edit box and clear the **Use default template** check box from the dialog box.
- Choose the **OK** button from the dialog box; the **New File Options** dialog box is displayed, as shown in Figure-29.

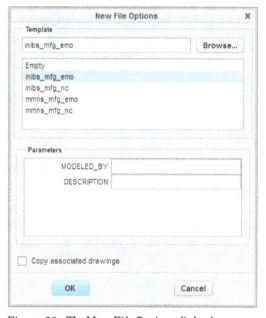

Figure-29. The New File Options dialog box

- Select the **mmns mfg nc** option from the list of templates displayed and choose the **OK** button from the dialog box. The Expert Machinist environment of the Creo Parametric will be displayed with metric units selected.

Creating a new NC Model

- Click on the **Create Model** tool from the **NC Model** panel of **Machining** tab in the **Ribbon**. A message box will be displayed as shown in Figure-30 and you will be prompted to enter the name of NC model.

Enter new NC Model name [C03_TUTORIAL1_NCMDL] :

Figure-30. The message box displayed

- Enter the name as **nc tutorial1** in the message box and choose the ✓ button from the message box. The **Open** dialog box will be displayed along with **NC MODEL** menu manager.
- Browse to the location **Creo Manufacturing\Chapter 3\tutorial**, select the c03 tutorial1.prt file and click on the **Open** button from the dialog box; the application window will be displayed, as shown in Figure-31.

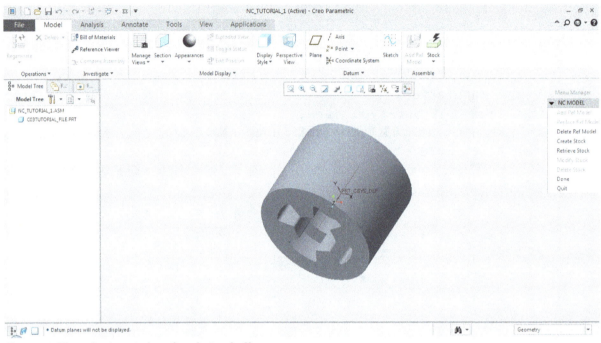

Figure-31. The application window after selecting the file

- Choose the **Retrieve Stock** option from the **NC MODEL** menu manager; the **Open** dialog box will be displayed again.
- Select the c03 tutorial file2.prt from the **Creo Manufacturing 4.0\ Chapter 4\tutorial** location and click on the **OK** button from the dialog box; the workpiece and reference model will be displayed, as shown in Figure-32.

Figure-32. The workpiece with the reference model

Now, we will constrain the workpiece with the reference model.

- Select the bottom faces of both the components. The coincident constrain will be applied automatically; refer to Figure-33. If not applied automatically, then choose the **Coincident** option from the **Constraint Type** drop-down in the **Placement** tab of the **Ribbon.**

Figure-33. The faces to be selected

- Select the **New Constraint** option from the **Placement** tab. You will be prompted to select a reference.
- Select the central axes of both the components and click on the **Coincident** option from the **Constraint Type** drop-down in the **Placement** tab. The components will be displayed as shown in Figure-34.

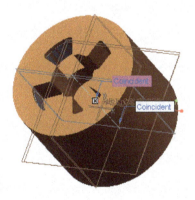

Figure-34. The workpiece assembled the reference model

- Click on the **Done** button from the **Ribbon** ✓ and choose the **Done** option from the **NC MODEL** Menu Manager. The Expert Machinist environment will be displayed along with the **Component Placement** contextual tab selected. You will be prompted to place the workpiece with reference part in Machine coordinate system.
- Click on the **Default** option from the **Component Placement** drop-down list in the **Ribbon** and click the **Done** button.

Analyzing for tool setup

Now, before we start creating a tool setup; it is required to check a few parameters of the workpiece like minimum radius, shortest edge of the workpiece, total depth of cut, material of the workpiece and so on.

Checking for minimum radius

• To check for minimum radius, click on the **Radius** tool from the **Measure** drop-down in the **Measure** group of the **Analysis** tab in the **Ribbon**; the Radius dialog box will be displayed, as shown in Figure-35.

Figure-35. The Radius dialog box

• Select the C03TUTORIAL FILE part from the **Model Tree** in the left of the application window; the minimum outside and inside radius will be displayed in the dialog box.

              Minimum Inside radius = -3
              Minimum Outside radius = 4

• On a close look at the model, you can see two type of arrows in the model. One type of arrows are green in color and other are blue in color. The green arrows represent the outside radius and blue arrows represent the inside radius.
• All the radius are displayed on the walls of the model. The walls are cut by side edges of the milling tool. So, there is no need of fillet at the bottom of the tool. Now, moving back to the tools available in Expert Machinist, you can find that **End Mill** is best for this machining as it does not have fillet at bottom.

Till this step, the tool is finalized as End Mill. Now, you will check for the diameter of the tool.

Checking for shortest edge

• To check for the shortest edge, choose the **Short Edge** tool from the **Model Report** group in the **Analysis** tab of the **Ribbon**; the **Short Edge** dialog box will be displayed, as shown in Figure-36.

Figure-36. The Short Edge dialog box

- Select the C03TUTORIAL FILE part from the Model Tree in the left of the application window, a message will be displayed in the dialog box as **16 edges shorter than 5.12250.**
- As per the analysis, the shortest edge is of 5.1225 mm. So, the only criteria here is radius in the reference part. It is already analyzed that minimum radius is 3. So, the diameter of the tool can be finalized as 6.

Indepth

The above analysis made for short edge might be confusing to some of my friends. Let's solve this mystery.

- Figure-37 shows a part with something like my company name engraved on it. If you are asked to define the size of end mill that can make those engraving cuts, what should be the criteria for diameter. Answer is; the shortest distance in the two edges of the engraving. Now, have a close look on the model in the Top view, refer to Figure-38. The areas of the part that might have shortest distance are highlighted in red boundary.
- Now, this is a time to try new tool i.e. **Distance**. Click on the **Distance** tool from the **Measure** drop-down in the **Measure** group of the **Analysis** tab in the **Ribbon**; you will be prompted to select a reference. Select the two point while holding the **CTRL** key, refer to Figure-39; the distance value will be displayed in a box as 1.48952 inch.

Figure-37. A part with engraving

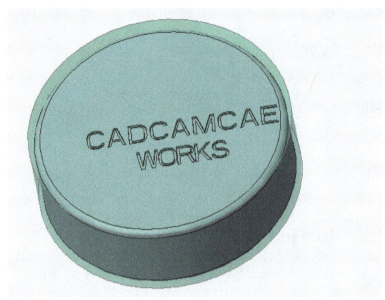

Figure-38. The areas of engraving expected to have shortest distance high-lighted in red

Similarly select the points and edges in **M and O,** refer to Figure-40.

Figure-39. Minimum distance value in A

Figure-40. Minimum distance value in M and O

Checking for Maximum depth in workpiece

Getting back to our tutorial, we need to check the maximum depth of cut from the top face so that we can avoid any accident caused by tool holder colliding with the workpiece. For example, if we take a drill having less length than the cutting depth required then the tool holder will get collided with the face in which the drilling is being done. You can check the maximum depth of cut by using the **Distance** tool.

• Click on the **Distance** tool from the **Measure** drop-down in the **Measure** group of the **Analysis** tab in the **Ribbon**; the **Measure: Distance** tool bar will be displayed and you will be prompted to select two references to find out the distance.

1. Select the two references to be selected, refer to Figure-41; the distance between the two edges will be displayed, refer to Figure-42.

Figure-41. The references to be selected

Figure-42. The distance value displayed between the edges

Till this step, we have finalized end mill as our tool. We calculated the total diameter of tool as 6.0 mm and total length of tool as 61.0 mm. Now, we will create tool setup for our desired tool. Note that also need to check the catalog of our tool supplier if such tools exist.

Creating tool setup

- Click on the **Machine Tool Manager** tool from the **Work Center** drop-down in the **Setup** group of the **Machining** tab of the **Ribbon**. The **Milling Work Center** dialog box will be displayed; refer to Figure-43.
- Select the **Tools** button from the **Tools** tab in the dialog box. The **Tools Setup** dialog box will be displayed, as shown in Figure-44.

Figure-43. Milling Work Center dialog box with Tools tab selected

Figure-44. The Tools Setup dialog box

- Specify the name as **End mill rough** in the **Name** edit box in the **General** tab of the dialog box.
- Click on the **End Mill** option from the **Type** drop-down, the dialog box will be displayed; as shown in Figure-45.

Figure-45. Modified **Tool Setup** *dialog box*

- Specify unit as Millimeter in the **Units** drop-down.

Now, we need to specify the values for tool geometry according to the value found by the analyses.

- Specify the values of geometry as shown in Figure-46.

Figure-46. The tool geometry with required values

Now, we will specify other parameters of the tool.

- Click on the **Settings** tab and specify the **Offset Number** as **01**.
- Click on the **Cut Data** tab and select the **Roughing** option from the **Application** drop-down list.
- Select the **Metric Unit System** radio button from the **Properties** area.

- Specify the **2500 rev/min**, **15 mm/min**, **1 mm**, and **1 mm** in the **Speed**, **Feed**, **Axial Depth**, and **Radial Depth** edit boxes in the **Cutting Data** area of the dialog box.
- Select the **ON**, **MEDIUM**, and **CW** options from the **Coolant Option**, **Coolant Pressure**, and **Spindle Direction** drop-down lists, respectively.
- Choose the **Apply** button; the tool will be displayed in the tool list of the dialog box.

Similarly, create a tool setup having following parameters:-
- Tool Name : End Mill Finish
- Tool Type : END MILL
- Tool Number : 2
- Offset Number : 02
- Application : Finishing
- Speed : 2000 rev/min
- Feed : 12 mm/min
- Axial Depth : 0.4 mm
- Radial Depth : 0.4 mm
- Coolant conditions same as above.

- Save the file by choosing the **Save tool parameters to a file** tool from the toolbar in the **Tools Setup** dialog box.
- Choose the **OK** button from the dialog box to exit.
- Choose the **Done** button from the **Milling Work Center** dialog box to exit.

PRACTICE 1

In this problem, you will import a reference model, as shown in Figure-47 and you will create a workpiece having uniform thickness of 2 over the reference model. You will create a tool setup for various operations on the workpiece. The model file for reference model is available in the Resources kit.

Figure-47. Reference model for Practice 1

PRACTICE 2

In this problem, you will import a reference model, as shown in Figure-48 and Figure-49. You will create a tool setup for various operations on the workpiece. The model file for reference model is available in the Resources kit.

Figure–48. Solid model for Practice 2

Figure–49. Refrence model for Practice 2

Chapter 4

Machine Setting

Topics Covered

The major topics covered in this chapter are:

- *Basics of Machine*
- *Use of 3 Axes, 4 Axes and 5 Axes machines*
- *Machines in Creo Manufacturing*
- *Setting Milling Work Center*

MILLING MACHINE

Machine is a component of manufacturing process that enables any cutting operation on the workpiece with the help of a tool. The types of machines have already been discussed. In Expert Machinist, there are three type of machines available i.e. 3-Axis, 4-Axis, and 5-Axis. In this chapter, you will learn about the applications and the settings of these machines.

3-AXIS MACHINE

In a 3-Axis machine, cutting operation can be performed in three directions; along X-axis, along Y-axis, and along Z-axis. Figure-1 illustrates a model of 3-Axis machine. There can be other type of 3-Axis milling machines having different directions. The arrows in the figure show the direction in which the cutting operation can be performed. In NC programming, three parameters will be required to specify the position of the tool while cutting.

Figure-1. Model of a 3-Axis milling machine

4-AXIS MACHINE

In a 4-Axis machine, cutting operation can be performed in three translational directions and one rotational direction. Figure-2 illustrates a model of 3-Axis machine. There can be other type of 4-Axis milling machines having different directions. The arrows in the figure show the direction in which the cutting operation can be performed. In NC programming, four parameters will be required to specify the position of the tool while cutting.

Figure-2. Model of a 4-Axis milling machine

5-AXIS MACHINE

In a 5-Axis machine, cutting operation can be performed in three translational directions and two rotational direction. Figure-3 illustrates a model of 5-Axis machine. The arrows in the figure show the direction in which the cutting operation can be performed. In NC programming, five parameters will be required to specify the position of the tool while cutting.

Figure-3. Model of a 5-Axis milling machine

USE OF 3-AXES, 4-AXES, AND 5-AXES MACHINES

Uses of different type of milling machines is mainly dependent on the job to machined. If machining is to be done in perpendicular to the workpiece then 3-Axes machine can do the job. For example the model given in Problem 1 of Chapter 3 can be machined easily by using the 3-Axes machine, refer to Figure-4. If the part to be machined is having some inclined pockets; refer to Figure-5, then you need to use the 4-Axes machine.

Figure-4. Model that can be machined by 3-Axes machine

Figure-5. Model that can be machined by 4-Axes machine

Similarly, if the part has an inclined cut in two or more directions then it can be machined by 5-Axes machine, refer to Figure-6.

Figure-6. Model that can be machined by 5-Axes machine

MACHINES IN CREO MANUFACTURING

In Creo Manufacturing application, you can generate the program codes for all the three types of milling machines discussed above. The procedure to add machine in Expert Machinist is given below.

- Click on the **Machine Tool Manager** from the **Work Center** drop-down in the **Setup** panel of the **Machining** tab in the **Ribbon**; refer to Figure-7. The **Milling Work Center** dialog box will be displayed as shown in Figure-8.

Figure-7. Machine Tool Manager

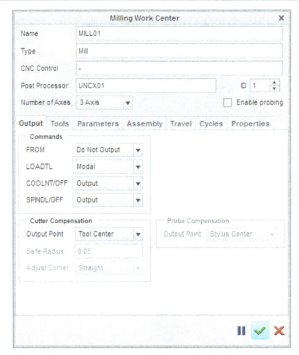

Figure-8. The Milling Work Center dialog box

In this dialog box, you can specify all the parameters related to your machine. These parameters are directly involved in the machining program being created by this software. Various parameters in this dialog box are discussed next.

Name

The **Name** edit box is used to specify a user defined name of the machine. This name is used to identify the machine when we are creating operation sequence and are prompted to select a machine. Also, the information related to machine is saved in *machinename.cel* format.

Type

The **Type** edit box displays the type of machine that you are using to perform operations. In case of Expert Machinist, it will remain **Mill**.

CNC Control

The **CNC Control** edit box is used to specify the name of CNC Controller, you are using in your machine, like Fanuc, Haas, Mazak, Siemens and so on. This field is optional to specify. But its good to specify full information because it will be reflected in the documentation of the program.

Post Processor

In **Post Processor** edit box, you can specify the name of post processor to be used for this machine. There is long list of post processor available with Creo. Post Processor is a kind of small program developed to create NC codes for the specific type of Machine. For example, Fanuc 15MA has a different post processor and HaasVF8 has a different post processor. You will learn about post processors later.

ID

This is a spinner as well as an edit box. You can specify the post processor id in this spinner/edit box. The value can be between 1-99.

Number of Axes

This option is a kind of drop-down. In this drop-down, there are three options available. Using these options, you can specify the number of axes available in your machine. You can select 3 Axis, 4 Axis, or 5 Axis machine.

Enable Probing

The **Enable Probing** check box is used to enable probing. Probing is the system that is used to find the position of workpiece with respect to tool position.

In spite of these common options, there are some other important options available in the tabs in the dialog box. These tabs and the practical meaning of the options in these tabs are discussed next.

Output tab

The options in the **Output tab** are used to specify the data related to the machine that is added in the output file.

FROM drop-down

The options in **From** drop-down are used to specify, whether the starting position of the tool before doing any cutting operation will be added in the output file or not. There are three options available in this drop-down list; **Do Not Output, Only at Start,** and **At Every Sequence.**

• Select the **Do Not Output** option, if you do not want to display the starting position of the tool in the output file.
• Select the **Only at Start** option, if you want to display the starting position of the tool at the beginning in the output file.
• Select the **At Every Sequence** option, if you want to display the starting position of the tool for every operation sequence in the output file.

LOADTL drop-down

The options of **LOADTL** drop-down are used to specify how the LOADTL command will work for the selected machine. LOADTL command is used to load tool from the Turret. There are three options in this drop-down; **Modal, Not Modal,** and **Not Modal on Position.**

• If the **Modal** option is selected then the LOADTL command will be added in the output file at only those locations where tool is being changed for a different operation.
• If the **Not Modal** option is selected then the LOADTL command will be added before every tool path sequence.

- If the **Not Modal on Position** is selected, then tool position will be created for each tool path sequence at starting as well as the end position of the tool path sequence.

COOLNT/OFF drop-down

The options OF **COOLNT/OFF** drop-down are used to specify whether the commands related to coolant will be reflected in the output file or the coolant operation will be controlled manually by machine panel.

- If you select the **Output** option, the **COOLNT/OFF** command will be added in the output file for each tool path sequence.
- If you select the **Do Not Output** option, the **COOLNT/OFF** command will be added only at the end of output file.

SPINDL/OFF drop-down

The options in the **SPINDL/OFF** drop-down are used in the same way as for **COOLNT/OFF**.

Output Point

The options in **Output Print** drop-down are used to the point that is considered as cutting end of the tool. There are two options available in this drop-down i.e. **Tool Center** and **Tool Edge**.

- If you select the **Tool Center** option from this drop-down, then the tool path will be created with reference to the tool center and will be reflected in the same way in the output file.
- If you select the **Tool Edge** option, then the tool path and codes in the output file will be generated according to the edge of tool.

Safe Radius

The **Safe Radius** edit box is available only when the **Tool Edge** option is selected in the **Output Point** drop-down. This edit box is used to specify the safe radius value that can be created by the selected tool on a concave corner. This value is dependent on the type of tool you are using for machining. You can find the safe radius value for your selected tool in the Tooling Catalogue of your tool supplier.

Adjust Corner

There are three options available in the **Adjust Corner** drop-down list. These options are used to specify the shape of workpiece at convex corners after the cutting operation is performed.

- If the **Straight** option is selected, then at the convex corners, the tool paths will be extended automatically until they intersect each other.
- If the **Fillet** option is selected, then fillet is automatically created at the convex corners in the tool path.
- The **Automatic** option is used when you want the system to automatically use the **Straight** or **Fillet** option.

Probe Compensation Output Point

The **Output Point** drop-down is active only when the **Enable Probing** check box is selected. The options in this drop-down are used to specify the point to be considered for generating the probing output file. There are two options available in this drop-down; **Stylus Center** and **Contact Point**.

- If you select the **Stylus Center** option, the output of probing will be generated according to the position of center of probe ball.
- If you select the **Contact Point** option, then the output of probing will be generated according to the position of contact point.

Rotation Area

The options of **Rotation Area** are available only if **4 Axis** is select in the **Number of Axes** drop-down list. The options in this area are discussed next.

Rotation Mode

The options in **Rotation Mode** drop-down are used to specify whether the value for rotation is according to MCS (Machine Coordinate System) or it incremental i.e. from the previous position of the tool.

Rotation Direction

The options in **Rotation Direction** drop-down are used to specify the direction of the tool rotation. By default, the **Shortest** option is selected in this drop-down. Hence the machine will use the shortest method to reach the specified angle, it can be CLW or CCLW. The other two options in this drop-down are used to specify the rotation method as clockwise and counter clockwise as per their names.

Rotation Axis

The **Rotation Axis** drop-down list is used to specify whether to use the A axis of the machine or B axis of the machine.

Tools tab

The options in **Tools** tab are used to specify the data related to tooling. There are two buttons and one spinner in this tab. The **Tools** button and its function have already been discussed. The other options are discussed next.

Tool Change Time (.sec)

This spinner is used to specify the time delay that is required by the machine to change the tool. If there are four tools in the turret named T1, T2, T3, and, T4; and you want to use the fourth tool after using the first tool. Let the tool change time is 2.5 seconds. Then the total time required to change from T1 to T4 will be :

```
Total Time     = Time mequired for(T1->T2 + T2->T3 + T3->T4)
               = 2.5 + 2.5 + 2.5 = 7.5 seconds.
```

Probes

The **Probes** button is active only when you select the **Enable Probing** check box in the **Milling Work Center** dialog box. This button is used to specify the setting related to probe for CMM validation.

- Click on **Probes** button from **Tools** tab. The **Probe Setup** dialog box will be displayed as shown in Figure-9. There are two options to add probes in the setup using this dialog box: either you can select from the Default Probes or you can use your own created probe.
- To use a default probe, click on the **File** button and select the **Default Probes** option from the flyout, the **Open** dialog box will be displayed as shown in Figure-10.

Figure-9. The Probe Setup dialog box

Figure-10. The Open dialog box displayed on selecting Default Probe option

- Select one of the default probe, the options related to the selected probe will be displayed, refer to Figure-11.

Figure-11. The Probe Setup dialog box after selecting default probe

The tabs and their options displayed in this dialog box are discussed next.

General Tab

The options in **General** tab are used to specify the general settings related to the probe. The options in this tab are dependent on the probe selected. Main parameters that can be specified by using these options are; number of tips of probe, roll angle, pitch angle and diameter of tip.

Settings Tab

The options in **Settings** tab are used to specify the probe number, register number, and comments related to the probe. Scanning check box in this tab is used to enable scanning of surface or segments that are generated automatically by measure points specified.

Probe Data Tab

The options in **Probe Data** tab are used to specify the feed rate at which the measurement will be done by the probe. You can specify measure feed in metric unit system as well as imperial unit system.

BOM Tab

The options in the **BOM** tab are used to specify the Bill of Material of the selected probe.

You will learn about Probes in more details in Chapter for CMM.

Parameters Tab

The options in the **Parameters** tab are used to specify parameters related to Power of machine and rotational speed of the spindle. The options of this tab are discussed next.

Maximum Speed (RPM)

The **Maximum speed(RPM)** edit box is used to specify the rotational speed of the spindle. The value specified in this edit box can not be negative. Also, the value specified will be in rounds per minute.

Horsepower

The **Horsepower** edit box is used to specify the power of main motor in HP unit.

Rapid Traverse

The options in **Rapid Traverse** drop-down are used to specify the unit to be used to specify the rapid traverse/feed rate.

Rapid Feed Rate

The **Rapid Feed Rate** edit box is used to specify the feed rate at which the tool will move in a non cutting step.

Defaults

The **Default** button is used to invoke the **SELECT SITE** Menu Manager; refer to Figure-12. The options in this Menu Manager are used to specify the location of SITE file(*.sit) for the machine. This file contains the parameters related to CMM which will be discussed later in this book.

Figure-12. SELECT SITE Menu Manager

- To link a SITE file with the Machine, click on **Current Dir** option from the Menu Manager. The **Open** dialog box will be displayed as shown in Figure-13.
- Select the desired file by using the browse option in dialog box and then choose the **Open** button. A message will be displayed at the information bar available at the bottom of the application window informing that the file has been successfully read; refer to Figure-14.

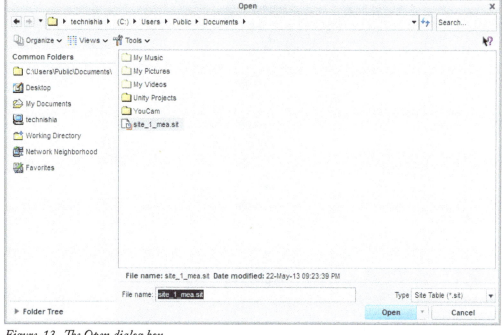

Figure-13. The Open dialog box

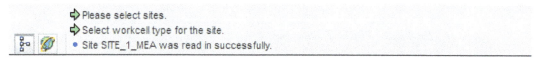

Figure-14. Message displayed in the information bar at the bottom of the application window

This message informs that the file has been linked successfully. Now, choose the **Resume** button { ▶ } in the **Milling Work Center** dialog box to activate the dialog box again.

PPRINT

The **PPRINT** button is used to invoke the **PPRINT** Menu Manager as shown in Figure-15. The options in this Menu Manager are used to specify the data that you want to include in the Cutter Location Data file(CL Data file).

- To specify the data to be included, click on the **Create** tool from the Menu Manager. The **Activate PPRINT** dialog box will be displayed; refer to Figure-16.

Figure-15. The PPRINT Menu Manager

Figure-16. The Activate PPRINT dialog box

- In this dialog box, various parameters are displayed in the Item column.
- In the **Option** column adjacent to desired parameter, you can specify the **Yes** or **No** option. The **Yes** option means the parameter will be included in the CL Data file. If you select the **No** option then the adjacent parameter will not be included in CL Data file.
- You can select multiple parameters at a time. After selecting the parameters, select the Yes button at the bottom of the dialog box.
- The selected parameters will be included in the CL Data file. You can select and deselect all the parameters by using the ▤ and the ▤ buttons, respectively.
- After specifying the parameters, click on the **OK** button from the dialog box. The **Modify**, **Save** and **Show** buttons will become active in the Menu Manager.

These options are discussed next.

Modify

If you select the **Modify** option, the active PPRINT file will be displayed in the **Activate PPRINT** dialog box. You can modify the values in the file by using the option in the dialog box.

Save

The **Save** option is used to save the PPRINT file. The procedure to save the file is discussed next.

- Click on the **Save** option. A message prompt will be displayed at the top of the modeling area of the application window and you will be prompted to specify the name of the file. Specify the name of the PPRINT file, refer to Figure-17.

Figure-17. The message prompt for PPRINT file name

- After specifying the name choose the green **Done** button. The file will be saved at the location displayed at the bottom of the information bar; refer to Figure-18.

Figure-18. The message displayed at the bottom after saving the file

Show

The **Show** option is used to display the content of active PPRINT file.

- Click on the **Show** option from the Menu Manager, the **INFORMATION WINDOW (pprinttbl.inf)** is displayed; refer to Figure-19. Note that you can not modify the content of the file by using this INFORMATION WINDOW.

Figure-19. INFORMATION WINDOW (pprinttble.inf)

Retrieve

The **Retrieve** option is used to link an already saved PRINT file with the current machine.

- On selecting the **Retrieve** option, the **Open** dialog box will be displayed.
- Select the desired file and click on **Open** button; the confirmation message will be displayed at the bottom in Information bar.

Assembly Tab

The options in the **Assembly** tab are used to add and link an already created machine assembly; refer to Figure-20. This assembly will also be used for the simulation purpose. The options in this tab are discussed next.

Figure-20. Assembly tab in the Milling Work Center dialog box

Machine Assembly

The **Machine Assembly** option is a type of information collector. This option displays the path of machine assembly to be added in the system.

- To add an assembly file, click on the **Open** button 🗁 adjacent to the collector. The **Open** dialog box will be displayed; refer to Figure-21.

Figure-21. The Open dialog box displayed on selecting Default Probe option

- Select the machine assembly file and click on the **Open** button from the dialog box. The assembly file name will be displayed in the **Machine Assembly** collector.

Coordinate System

The **Coordinate System** collector is active only after you have selected the machine assembly file. This collector is used to store the coordinate system according to which the machine assembly is to be placed. The coordinate system selected is directly matched with the coordinate system of the machine assembly while placing.

You can see the preview of the machine assembly by using the preview 👓 button.

Travel

The options of **Travel** tab are used to specify the total travel of tool that is available for the selected machine. The **Milling Work Center** dialog box after choosing this tab is displayed as shown in Figure-22.

Figure-22. The Travel tab in the Milling Work Center dialog box

- As you can see from the above figure, there are three edit boxes for each direction. You need to specify the value of total travel in any two of the three edit boxes for each direction, the last edit boxes will get the value accordingly, refer to Figure-23.

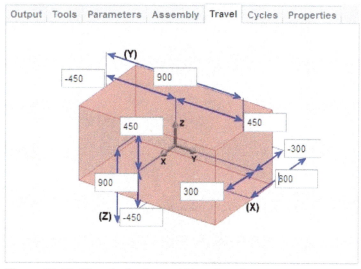

Figure-23. The Travel tab after specifying the values

Cycles

The options in the **Cycles** tab are used to configure hole making cycles for the machine. The **Milling Work Center** dialog box after choosing this tab is displayed, as shown in Figure-24.

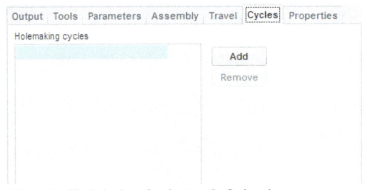

Figure-24. The dialog box after choosing the Cycles tab

- Here you can add as many number of hole making cycles as you want. These cycles will be stored with the machine and will be used later while creating program.
- To add a hole making cycle, click on the **Add** tool from the dialog box. The **Customize Cycle** dialog box will be displayed as shown in Figure-25.

Figure-25. The Customize Cycle dialog box

The options in this dialog box are discussed next.

Cycle Name

The **Cycle Name** edit box is used to specify the name of the hole making cycle.

Cycle Type

The **Cycle Type** edit box is used to specify a keyword for the current cycle.

Cycle Prompts

This area stores the information regarding the starting surface and ending surface of the cycle. Preview of the starting surface and end surface is displayed in the right of this area. If you right click on empty space in this area, a shortcut menu is displayed as shown in Figure-26.

Add Variable

Add Reference

Add Expression

Change Description

Redefine Expression

Remove Variable

Remove Reference

Remove Expression

Figure-26. Shortcut menu displayed

Using this shortcut menu, you can add options like variable, reference, or expression for the cycle. The values regarding the added option can be specified in **Name**, **Type**, and **Description** edit boxes. The options that are inactive gets activated after you select an expression, variable or reference. The **Customize Cycle** dialog box will be discussed in detail in later chapters.

Properties

The options in the **Properties** tab are used to edit and/or display properties of the created machine.

• Click on the **Properties** tab. The **Milling Work Center** dialog box is displayed as shown in Figure-27. The options in this tab are discussed next.

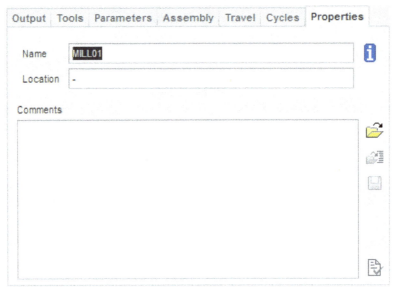

Figure-27. The dialog box after choosing the Properties tab

Name

The **Name** edit box is used to specify the name of the machine to be created. The information button adjacent to this edit box is used to display the complete information of the machine.

Location

The **Location** edit box is used to specify the location of machine files.

Comments

The **Comments** edit box is used to specify comments about the selected machine.

• You can also load comments from text files by using the **Open** 📂 button displayed adjacent to the edit box.
• Using the **Insert** 📝 button you can insert more text in the file.
• Using the **Save** 💾 button, you can save the loaded comment.
• The **Insert** and the **Save** buttons will be active only after specifying at least one character in the **Comments** edit box.

PRACTICAL 1

In this tutorial, you will create a machine with following parameters:

• Name of machine is **Fanuc milling center01** with post processor UNCX03 and number of axes = 3. CNC Control is Fanuc 18i.

- Coolant added in the output file
- From location to be added for each tool movement in the output file
- Tool corner to be filleted, radius for safe corner should be 0.8 mm from the tool edge.
- 3 tools with the specifications as shown in Figure-28 and changing time for tool should be 4 second.
- Maximum RPM of machine spindle is 7500 and rapid feed rate is 25 mm per minute.
- Total travel of machine is 900 mm in each direction.

Figure-28. Specifications for the tools

Starting Creo Expert Machinist

- Start Creo Parametric and choose the **New** button from the **Ribbon**.
- Select **Manufacturing** radio button from the **Type** area and **Expert Machinist** from the **Sub-type** area of the dialog box.
- Specify the name of the file as **C04 tut1** and clear the **Use default template** check box.
- Now, choose the **OK** button from the dialog box, the **New File Options** dialog box will be displayed. Select the **mmns mfg nc** template and then choose the **OK** button.

Adding an NC model in the system

- Click on the **Create Model** tool from **NC Model** panel in the **Ribbon**. An edit box will be displayed to define the name of the nc model.
- Choose the **Done** button ✅ from the prompt, the default name will be accepted and you will be prompted to select a part file as reference model.

- Select the file c2 tutorial1 by using the options in the dialog box and then choose the **Open** button.
- Click on the **Create Stock** option from the **Menu Manager** displayed in the right; refer to Figure-29. The **Auto Workpiece Creation** tab will be added in the **Ribbon** and the related options will be displayed.

Figure-29. Menu Manager

- Create the workpiece as per your requirement and click the **Done** button from the **Ribbon** and the **Menu Manager**, the **Component Placement** tab will be displayed.
- Place the workpiece at the default position and choose the **Done** button ✅ from the **Ribbon**.

Creating Machine Setup

- Click on the **Machine Tool Manager** tool from the **Work Center** drop-down in the **Setup** panel, the **Milling Work Center** dialog box will be displayed. Click in Name edit box displayed at the top and specify the name as **Fanuc milling center01**.
- Enter the name of Post processor as UNCX03 in the **Post Processor** edit box. Set the value of **Number of Axes** drop-down to 3 if not set by default.
- Enter **Fanuc 18i** in the **CNC Control** edit box.
- Click in the **COOLNT/OFF** drop-down and select **Output** option, if not selected.
- Select the **At Every Sequence** options from the **FROM** drop-down.
- Select the **Tool Edge** option from the **Output Point** drop-down, the options below the drop-down become active.
- Click in the **Safe Radius** edit box and specify the value as **0.8** mm. Also, select the **Fillet** option from the **Adjust Corner** drop-down.
- Click in on the **Tools** tab and select the **Tools** button, the **Tools Setup** dialog box will be displayed.
- Specify the parameters as displayed in Figure-28 (**a**) and then choose the **Apply** button displayed at the bottom. The tool will be added in the setup; refer to Figure-30.

Figure-30. The Tool Setup dialog box after adding a tool

- To add more tool, choose the **New** button displayed in the toolbar at the top and specify the parameters in the same way as in step 9.

Tip
You can also save the tools if you want to used them later.

- Choose the **OK** button after adding all the required tools, the Milling Work Center dialog box will be displayed again.
- Set the value **04** in the **Tool Change Time (sec.)** spinner.
- Select the **Parameters** tab in **Milling Work Center** dialog box, and enter **7500** in the **Maximum Speed (RPM)** edit box.
- Select **MMPM** in the **Rapid Traverse** drop-down, if not selected.
- Enter the value as **25** in the **Rapid Feed Rate** edit box.
- Select the **Travel** tab and specify the value of total travel 900 each direction i.e. 450 in positive and 450 in negative of each direction.
- Now, choose the **Done** button from the bottom to create the machine.

PRACTICE

In this problem, you will create a machine with following parameters:

1. Name of machine is **Haas Milling Center01** with post processor UNCX01 and number of axes = 4. CNC Control is Haas.
2. From position of tool should be displayed only at the start of operation.
3. Coolant added in the output file.
4. Tool corner to be filleted, radius for safe corner should be 0.8 mm from the tool edge.
5. 2 tools with the specifications as shown in Figure-31 and Figure-32, and changing time for tool should be 6 second.
6. Maximum RPM of machine spindle is 9000 and rapid feed rate is 250 mm per minute.

Total travel of machine is 900 mm in X direction, 600 in Z direction and 500 in Y direction.

Figure-31. Specification of first tool

Figure-32. Specification of second tool

Chapter 5

Operations and Cutting Strategies

Topics Covered

The major topics covered in this chapter are:

- *Machining Operations*
- *Operations in Creo Manufacturing*
- *Various cutting strategies for machining in Creo Manufacturing*
- *Setting Milling Work Center*

INTRODUCTION

In any kind of machining, operations are the cutting steps to get the required shape and size from the workpiece. In some cases, a single operation can cut only a small portion of the workpiece while in some cases, a single operation can cut almost all the workpiece to get the desired parameters in one go. Area of cut and depth of cut depends upon the type of operation to be performed and cutting tool selected. The function of operations in Creo Manufacturing is discussed next.

OPERATIONS IN CREO MANUFACTURING

In Creo Manufacturing, operation is a process that cuts material from the workpiece. An operation is composed of various cutting strategies. The operation can have more than one tools involved in the process. The process is discussed next.

- Click on the **Operation** tool from the **Setup** panel in the **Home** tab of the **Ribbon**. The **Operation** tab will displayed in the **Ribbon**; refer to Figure-1.

Figure-1. The Operation tab in the Ribbon

The options of this tab are discussed next.

Work Center

The options of **Work Center** drop-down are used to specify a machine for the selected operation. The machines created in the current session are displayed in this drop-down list.

Program Zero

This is a type of value collector. This option stores the coordinate system that is defined as the zero position for the current operation. To define a program zero position, click in this option and then select the coordinate system that you want to specify as the program zero position.

Parameters Tab

The options in the **Parameters** tab are used to specify the basic parameters regarding the cutter location(CL) file of current operation, refer to Figure-2. The options in this tab are discussed next.

Figure-2. The Parameters Tab

PARTNO

The **PARTNO** edit box is used to specify the part name or number. This name/number is displayed at the top of the nc program to be generated. This name/number is used to identify the nc programs later.

Startup File

The **Startup File** edit box is used to specify the name of the CL file that is to be included at the starting of nc program. These codes are added in the file after Part number, machine, and unit codes in the nc program. To add a startup file, copy the required file in the current working directory and then specify the name of the file in the **Startup File** edit box.

Shutdown File

The **Shutdown File** edit box is used to specify the name of the CL file that is to be included at the end of the nc program. This edit box works in the same way as **Startup File** edit box.

Output File

The **Output File** edit box is used to specify the name of the output file to be created for the current operation. To specify the names of the files in all the edit boxes, you should use alphanumeric keys only.

Options Tab

The option in the **Options** tab is used to specify the stock material for the current operation. If you have selected a 4 axes milling machine then the options related to rotation of tool will also be displayed in this tab, refer to Figure-3. The options in this tab are discussed next.

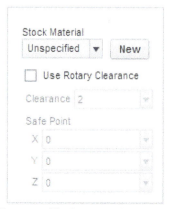

Figure-3. The Options tab

Stock Material

The options in the **Stock Material** drop-down list are used to specify the conditions of the stock material. The specified stock material name is saved in the mfg wp material list file in the default directory.

Use Rotary Clearance

The **Use Rotary Clearance** check box is used to enable safe clearance distance for the rotary bed of the 4 Axes machine. On selecting this check box, the options below this check box become active.

Clearance

The **Clearance** edit box is used to specify the clearance value for the tool when the table rotates. For example, the value is specified in this edit box as 5. Then the tool will move to such a position that it is away from table rotation range by 5 mm. Note that this rotation range also counts the fixtures and other assemblies attached to the table.

Safe Point

There are three edit boxes available under this head. Specify the safe values in X, Y, and Z directions for the tool at which the tool will retract when table rotate.

Fixture Setup

The options in the **Fixture Setup** tab are used to set a fixture for workpiece in the current operation. The process to add the fixture is given next.

• Click on the **Add a fixture component** button from the tab. The **Open** dialog box will be displayed and you will be prompted to select an assembly or a part for the fixture, refer to Figure-4.

Figure-4. The Open dialog box displayed on selecting Add a fixture Component button

- Click on the **Type** drop-down, if you want to select a file of other format.
- After selecting the desired format, browse and select the file you want to use as fixture and then click the **Open** button; the **Component Placement** tab will be displayed in the **Ribbon** and the selected fixture part will be displayed in the modeling area, refer to Figure-5.

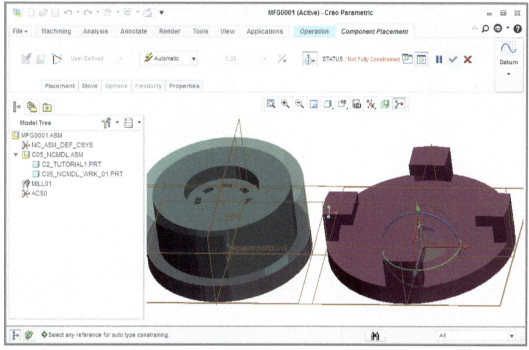

Figure-5. The Component Placement tab and the fixture part

- Assemble the Fixture and the workpiece as desired.
- Click on **Done** button from the **Ribbon**, the workpiece and the fixture get assembled and displayed in the modeling area and **Fixture** tab, refer to Figure-6.

Figure-6. The modified Operation tab and the assembled workpiece

Process

The options in the **Process** tab are used to specify actual time taken by a machining process. Also, you can find out the estimated time taken by any machining operation; refer to Figure-7.

- The **Calculated Time** is active only when you have generated tool path for the current operation.
- To calculate the estimated time for current operation, click on the **Recalculate Machining Time** button ⏰ from process tab. The estimated time will be displayed in the **Calculated Time** edit box; refer to Figure-8.

Figure-7. The Process tab

Figure-8. The Process tab with calculated values

You can find out the actual time taken by the machine and compare it with the calculated time to improve your machining process.

Properties

The options in the **Properties** tab are similar to the options discussed for **Properties** tab in **Milling Work Center** dialog box in chapter 4.

After creating an operation, the tools in the **NC Features & Machining** panel of the **Ribbon** get activated; refer to Figure-9. The panel and its tools are discussed next.

Figure-9. Panel for creating cutting strategies

NC FEATURES & MACHINING

The options in this panel are used to create cutting strategies which in turn can be used to create tool paths. These options and their patterns of creating tool paths are discussed next.

Face

As the name suggests, this option is used to perform facing operation. A facing operation is used to remove material from the workpiece to create flat face. The procedure to use this tool is discussed next.

- Click on the **Face** tool from the **NC Features & Machining** panel. The **Face Feature dialog** box along with the **SELECT SRFS** Menu Manager will be displayed; refer to Figure-10.

Figure-10. Face Feature dialog box with the SELECT SRFC Menu Manager

- By default, the **Define Feature Floor** button is selected; so you are prompted to select the feature floor.
- Boundaries of the selected feature floor will become boundaries of the face milling area.
- Figure-11 shows a workpiece assembled with the reference model and the face to be selected for face milling. Figure-12 shows the area that will be machined by face milling feature.

Figure-11. Workpiece and reference model assembled together; the green highlighted line shows the face to be selected

Figure-12. The material to be removed from the workpiece

- Now, choose the **Done/Return** option from the Menu Manager. The volume to be machined will be highlighted; refer to Figure-12.
- Click **OK** button from the **Face Feature** dialog box. The face milling feature will be created and displayed in the **Model Tree** at the left of the application window; refer to Figure-13.
- In the name displayed in Model Tree, OP010 represents the operation number under which this machining process is to be performed.

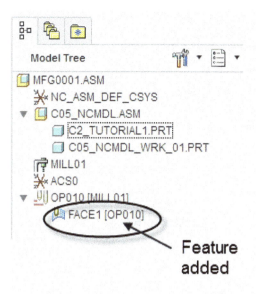

Figure-13. Face milling feature added

The options in of the **Face Feature** dialog box are discussed next.

Feature Name

The **Feature Name** edit box is used to specify the name for the face milling feature. The name can be specified in alphanumeric letters as well as special letters.

Define Feature Floor

The **Define Feature Floor** option is used to select the feature face that is to be machined.

- To select the feature face, choose the arrow button ![cursor] adjacent to this option, the **SELECT SRFS** Menu Manager will be displayed. If you have already selected the feature face then the **Warning** message box will be displayed; refer to Figure-14.
- It warns that the selection can affect the current selections.
- Choose the **Continue** button in such case. The **SELECT SRFS** Menu Manager will be displayed. The options in the Menu Manager are as follows:

Add

Select the **Add** option from the Menu Manager if you want to add more faces for face milling and then click on the surfaces that you want to add.

Remove

Select the **Remove** option if you want to remove a surface from the current selection and then select the surfaces that you want to remove from the selection. This option is active only when you already have some selected faces.

Figure-14. The Warning message box

Show

The **Show** option is used to highlight the faces that are selected for the face milling process.

Done/Return

The **Done/Return** option is used to accept the selections and return to the Face Feature dialog box.

Define Program Zero

The **Define Program Zero** option is used to define the zero position of the program. The procedure to use this option is discussed next.

- Click the arrow button ⬆ adjacent to this option. The **WIND CSYS** Menu Manager will be displayed; refer to Figure-15.
- You are prompted to select a coordinate system. Select the coordinate system where you want to define the program zero position, the Menu Manager will disappear automatically.

Figure-15. The WIND CSYS
Menu Manager

The ⚭ button is used to display the preview for the adjacent option.

Adjust Feature Bottom

The **Adjust Feature Bottom** option is used to change the depth of face milling. The procedure is discussed next.

- To change the total depth of cut for face milling, click on the arrow button adjacent to this option. The **Define/Adjust Feature Bottom** dialog box will be displayed as shown in Figure-16.

The options in this dialog box are discussed next.

Figure-16. The Define/Adjust Feature Bottom dialog box

Offset Above floor

Select this radio button if you want to specify the maximum depth of cut above the current selected face. Specify the distance value in the edit box adjacent to this radio button by which you want to shift the cutting depth. A negative value in this edit box will move the maximum depth of cut to below the current selected face.

Use Datum Plane

Select this radio button if you want to select an already existing datum plane as the maximum depth of cut. Click on the arrow button and selected the desired datum plane.

Offset from Program Zero

Select this radio button if you want to specify the maximum depth of cut above or below the program zero by a specified value. Specify the distance value in the edit box adjacent to this edit box.

None

The **None** option is used to remove any override applied on the maximum depth of cut.

Choose the **OK** button from the dialog box to apply the changes.

In short the Face operation is used to remove material from the face of the workpiece. This should be the first operation to be performed on flat face parts.

Profile

As the name suggests, this milling operation is used to cut material as per the profile selected. Note that the profile can only be selected on side walls of the workpiece. The procedure to use this tool is discussed next.

• Click on the **Profile** tool from the **NC Features & Machining** panel of the **Machining** tab in the **Ribbon**. The **Profile Feature** dialog box will be displayed along with **SELECT SRFS** Menu Manager and **Select** dialog box; refer to Figure-17.

Figure-17. Profile Feature dialog box with SELECT SRFS Menu Manager and Select dialog box

- Also, you will be prompted to select a surface (check the Information Bar displayed at bottom by dragging the splitting line upwards).
- Select the side walls of the reference model to remove the material from sides of the workpiece as per the selected surface; refer to Figure-18.

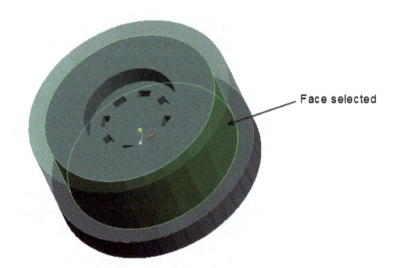

Figure-18. Face selected for Profile milling

- Click on the **Done/Return** button from the Menu Manager. In some cases after selecting the **Done/Return** button, you may get the **Mill Feature Error** message box; refer to Figure-19.

Figure-19. Mill Feature Error message box

- In such cases, you will be prompted to adjust the boundaries of the milling area that is created after selecting the feature.
- Choose the **Adjust** button from the dialog box, top view of the selected surfaces will be displayed and the Sketcher environment will be invoked automatically; refer to Figure-20.

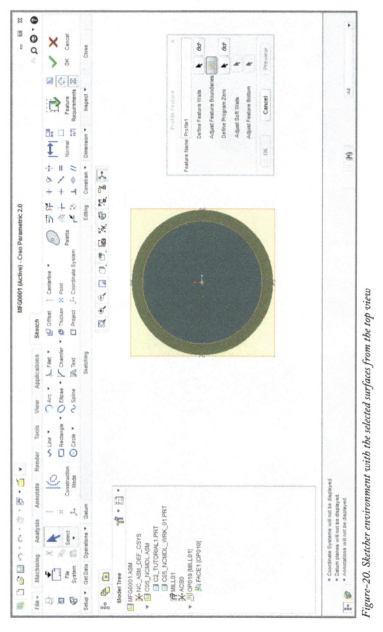

Figure-20. Sketcher environment with the selected surfaces from the top view

- From the figure, you can see that the outer boundaries created automatically are rectangular. But the boundaries of our workpiece are circular. So, we need to modify the boundaries so that it looks like Figure-21.

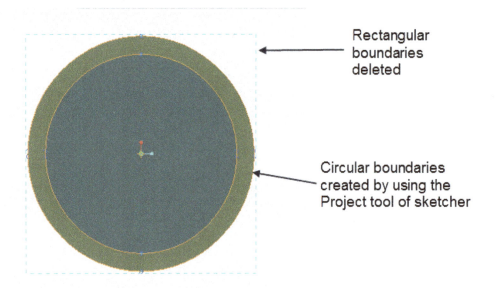

Rectangular
boundaries
deleted

Circular boundaries
created by using the
Project tool of sketcher

Figure-21. Selected surfaces after modifications

• After modifying the boundaries, choose the **OK** button from the **Ribbon**. The preview of workpiece with the milling volume will be displayed automatically; refer to Figure-22.

Figure-22. Preview of the workpiece

• Click on the **OK** button from the **Profile Feature** dialog box to create the operation.

The options displayed in the **Profile Feature** dialog box are discussed next.

Define Feature Walls

The **Define Feature Walls** option is used to select the walls on which the profile milling is to be performed. Figure-23 shows a workpiece with reference model having curves on the walls.

- To machine those walls, click on the arrow button adjacent to **Define Feature Walls** option, **SELECT SRFS** Menu Manager will be displayed and you will be prompted to select the wall surfaces for profile milling.

Figure-23. Walls to be selected for Profile Milling

- Select the wall faces; refer to Figure-23 and then choose the **Done/ Return** option from the Menu Manager. Note that the selected surfaces must be consecutive. If your selection is right then all the buttons in the **Profile Feature** dialog box will become active.

Adjust Feature Boundaries

The **Adjust Feature Boundaries** option is used to adjust the boundaries for the milling area. The procedure to use this tool is discussed next.

- Click on **Adjust Feature Boundaries** option, the sketcher environment is invoked automatically.
- Using the sketcher tools, you can modify the area to be machined by the operation. For example, in the above case, we can specify a small belt as milling area for the operation; refer to Figure-24.
- Figure also shows the output that will be generated after making such changes.
- After modifications in the sketcher, click on the **OK** button from the **Ribbon,** the output volume will be displayed. If you have created a wrong sketch then some of the buttons in the **Profile Feature** dialog box will become inactive and you need to modify the sketch again.

Figure-24. Left figure shows the sketch created after modifications in Sketcher environment Right figure shows the milling volume that will be machined

Define Program Zero

The **Define Program Zero** button has the same function for each of the milling operations. This option has already been discussed.

Adjust Soft Wall

The **Adjust Soft Wall** option is used to convert the Hard Walls into Soft Walls. In Creo Manufacturing, any wall or face that belongs to workpiece is termed as Soft. Whereas any wall or face that belongs to the reference model is termed as Hard. So, using this option, you can transfer any face to soft wall. Hard walls define the boundaries of the output model so the tool path around the hard walls are more accurate and dense with respect to the soft walls. The procedure to use this option is discussed next.

- To add a hard wall to soft wall, click on the **Adjust Soft Wall** option; the walls of selected surfaces will be highlighted in violet color, refer to Figure-25.

Figure-25. Walls highlighted in violet color

- Also the **SELECT SRFS** Menu Manager will be displayed. Select the surfaces that you want to add in the soft walls, refer to Figure-26.
- After selecting the walls, choose the **Done/Return** button from **SELECT SRFS** Menu Manager; soft walls will be created.
- Note that when you choose the Preview button from the **Profile Feature** dialog box, the soft walls are displayed in CYAN color and hard walls are displayed in RED color.

Figure-26. The highlighted faces to be transferred to soft walls

Adjust Feature Bottom

The **Adjust Feature** option is discussed earlier in **Face** option.

After defining all the options, click on the **OK** button from the dialog box to create the feature.

Step

The **Step** tool is used to create steps in the milling process. The step milling feature is used where you want to create pins or other boss features on the workpiece; refer to Figure-27. The procedure to use this tool is discussed next.

- Click on the **Step** tool from the **NC Feature & Machining** panel of the **Machining** tab in the **Ribbon**. The **Step Feature** dialog box will be displayed along with the **SELECT SRFS** Menu Manager; refer to Figure-28.

Figure-27. Figure 5-27 Workpiece with steps to be machined

Figure-28. The Step Feature dialog box with the SELECT SRFS Menu Manager

- You will be prompted to select surface to define base of the step feature. Select the surface as shown in Figure-29.
- Click on the **Done/Return** option from the Menu Manager. If the selection is right, preview of the feature will be displayed automatically; refer to Figure-30.

Figure-29. Surface to be selected for the step feature

- In this figure, the red boundaries show the hard walls and cyan boundaries show the soft walls. Choose the **Adjust Soft Walls** option if you want to add any wall to soft walls. Like in this case, a wall is desired to be in soft side; refer to Figure-31.

Figure-30. Preview of the Step feature

Figure-31. Walls after shifting

- Click on the **OK** button from the dialog box to create the feature.

The options in this dialog have already been discussed.

Slot

As the name suggests, the **Slot** tool is used to create slots in the workpiece. The process to use this tool is discussed next.

- Click on the down arrow displayed adjacent to **Slot** in the **NC Features & Machining** panel of the **Ribbon**. A list of tools will be displayed.
- Now, click on **Slot** in this list to activate the tool. The **Slot Feature** dialog box is displayed along with the **SELECT SRFS** Menu Manager; refer to Figure-32.
- Select the surfaces that you want to include for the slot feature; refer to Figure-33.

Figure-32. Slot Feature dialog box with SELECT SRFS Menu Manager

Figure-33. Face to be selected for slot feature

- Now, click on the **Done/Return** button from the Menu Manager. The preview of the slot feature will be displayed; refer to Figure-34.

Figure-34. Preview of slot feature

- Choose the **OK** button from the **Slot Feature** dialog box to create the feature.

The options in this dialog box have already been discussed in this chapter.

Through Slot

As the name suggests, this tool is used to create a through slots in the workpiece. The procedure to use this tool is discussed next.

- Click on the down arrow displayed adjacent to **Slot** in the **NC Features & Machining** panel of the **Ribbon**. A list of tools will be displayed.
- Now, click on **Through Slot** in this list to activate the tool. On choosing this tool, the **Through Slot Feature** dialog box is displayed along with the **SELECT SRFS** Menu Manager; refer to Figure-35.
- Also, you are prompted to select walls of the reference model where you want to create the through slot feature.

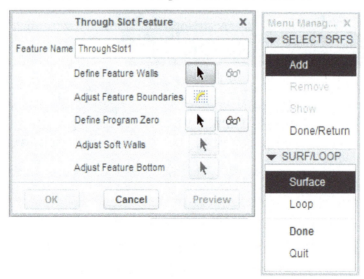

Figure-35. The Through Slot Feature dialog box with the SELECT SRFS Menu Manager

- Select the walls of the reference model; refer to Figure-36 and click on the **Done/Return** button from the **Menu Manager.** Preview of the feature will be displayed; refer to Figure-37.

Figure-36. Model in wireframe and shading style with the faces to be selected

Figure-37. Preview of the through slot feature

- Click on the **OK** button from the **Through Slot Feature** dialog box. The slot feature will be created and displayed in the **Model Tree.**

The options available in the **Through Slot Feature** dialog box have already been discussed.

Channel

As the name suggests, the **Channel** tool is used to create channel of two more slots, intersecting each other, in the workpiece. The procedure to use this tool is discussed next.

• Click on the down arrow displayed adjacent to **Slot** in the **NC Features & Machining** panel of the **Ribbon**, a list of tools will be displayed.
• Click on **Channel** in this list to activate the tool. The **Channel Feature** dialog box is displayed along with the **SELECT SRFS** Menu Manager; refer to Figure-38. Also, you will be prompted to select floor of the reference model where you want to create the channel feature.

Figure-38. Channel Feature dialog box with SELECT SRFS Menu Manager

• Select the surface; refer to Figure-39 and click the **Done/Return** option from the Menu Manager. Preview of the channel feature will be displayed; refer to Figure-40.

Figure-39. Model with the face to be selected for slot feature

Figure-40. Preview of the slot feature

- Choose the **OK** button from the dialog box to create the feature. The options in the **Channel** dialog box are same as discussed earlier.

Pocket

As the name suggests, the **Pocket** tool is used to remove material from the workpiece in the form of a pocket. This tool is used where we have the slot bound by hard walls. The procedure to use this tool is discussed next.

- Click on the down arrow displayed adjacent to **Pocket** in the **NC Features & Machining** panel of the **Ribbon**, a list of tools will be displayed.
- Now, click on **Pocket** in this list to activate the tool. The **Pocket Feature** dialog box will be displayed along with the **SELECT SRFS** Menu Manager; refer to Figure-41. Also, you will be prompted to select floor of the reference model where you want to create the pocket feature.

Figure-41. Pocket Feature with SELECT SRFS Menu Manager

- Select the floor surface, refer to Figure-42 and click the **Done/Return** option from the Menu Manager. The preview of the pocket feature will be displayed refer to Figure-43.

Face to be selected

Figure-42. Face to be selected for pocket feature

Figure-43. Preview of the pocket feature

- Choose the **OK** button from the **Pocket Feature** dialog box to create the pocket feature. The options of the **Pocket Feature** dialog box are same as discussed earlier.

Through Pocket

The **Through Pocket** tool works in the same way as **Pocket** tool does with the only difference that there is no floor. This tool is used where we have a through slot bound by hard walls. The procedure to use this tool is discussed next.

- Click on the down arrow displayed adjacent to **Pocket** in the **NC Features & Machining** panel of the **Ribbon**. A list of tools will be displayed.
- Now, click on **Through Pocket** in this list to activate the tool. The **Through Pocket Feature** dialog box will be displayed along with the **SELECT SRFS** Menu Manager; refer to Figure-44. Also, you will be prompted to select walls of the reference model where you want to create the through pocket feature.

Figure-44. Through Pocket Feature dialog box with the SELECT SRFS Menu Manager

• Select the wall surfaces; refer to Figure-45 and choose the **Done/ Return** button Menu Manager. The preview of the through pocket feature will be displayed; refer to Figure-46.

Figure-45. Walls to be selected for through pocket feature

Figure-46. Preview of the through pocket feature

- Choose the **OK** button from the dialog box to create the feature, the options in the **Through Pocket Feature** dialog box have already been discussed.

Flange

The **Flange** tool is used to remove large amount of material from the face of workpiece. The procedure to use this tool is discussed next.

- Click on the down arrow displayed adjacent to **Flange** in the **NC Features & Machining** panel of the **Ribbon**. A list of tools will be displayed.
- Now, click on **Flange** in this list to activate the tool. The **Flange Feature** dialog box is displayed along with the **SELECT SRFS** Menu Manager; refer to Figure-47. Also, you will be prompted to select floor of the reference model where you want to create the flange feature.

Figure-47. Flange Feature dialog box with the SELECT SRFS Menu Manager

- Select the face of the reference model; refer to Figure-48 and then click on the **Done/Return** button. Preview of the flange feature will be displayed; refer to Figure-49.

Figure-48. Face to be selected for flange feature

Figure-49. Preview of the flange feature

Slab

The **Slab** tool is used to remove large amount of material from the face of workpiece. The procedure to use this tool is discussed next.

*   Click on the down arrow displayed adjacent to **Flange** in the **NC Features & Machining** panel of the **Ribbon**. A list of tools will be displayed.
*   Now, click on **Slab** in this list to activate the tool. The **Slab Feature** dialog box is displayed along with the **SELECT SRFS** Menu Manager; refer to Figure-50. Also, you will be prompted to select floor of the reference model where you want to create the slab feature.

Figure-50. Slab Feature dialog box with SELECT SRFS Menu Manager

*   Select face of the model on which you want to perform slab milling, refer to Figure-51 and click the **Done/Return** option from the Menu Manager. Preview of the slab feature will be displayed; refer to Figure-52.

Figure-51. Face selected for slab milling

Figure-52. Preview of the slab feature

Note that in the preview, the faces coplanar to the selected face will be machined automatically. The options in the **Slab Feature** dialog box are the same as discussed earlier.

Top Chamfer

The **Top Chamfer** tool is used to chamfer on the top face of workpiece. The procedure to use this tool is discussed next.

- To access this tool, click on the down arrow displayed adjacent to **Top Chamfer** in the **NC Features & Machining** panel of the **Ribbon**. A list of tools will be displayed.
- Now, click on **Top Chamfer** in this list to activate the tool. The **TOP CHAMFER** dialog box will be displayed along with the **SELECT SRFS** Menu Manager; refer to Figure-53. Also, you will be prompted to select floor of the reference model where you want to create the chamfer feature.

Figure-53. TOP CHAMFER dialog box with the SELECT SRFS Menu Manager

- Select the chamfered face as shown in Figure-54 and choose the **Done/Return** button from the Menu Manager. Preview of the chamfer feature will be displayed as shown in Figure-55.

Figure-54. Chamfered faces to be selected

Figure-55. Preview of the chamfer feature

- Choose the **OK** button from the dialog box to created the feature.

Top Round

The **Top Round** tool is used to create round on the top face of workpiece. The procedure to use this tool is discussed next.

- To access this tool, click on the down arrow displayed adjacent to **Top Chamfer** in the **NC Features & Machining** panel of the **Ribbon**. A list of tools will be displayed.
- Now, click on **Top Round** in this list to activate the tool. On choosing this tool, the **TOP ROUND** dialog box will be displayed along with the **SELECT SRFS** Menu Manager; refer to Figure-56.
- Also, you will be prompted to select floor of the reference model where you want to create the round feature.

Figure-56. TOP ROUND and SELECT SRFS Menu Manager

This tool and its options work in similar way as the **Top Chamfer** tool do.

Ribtop

The **Ribtop** tool is used to machine the top face of a rib. The procedure to use this tool is discussed next.

- To access this tool, click on the down arrow displayed adjacent to **Ribtop** in the **NC Features & Machining** panel of the **Ribbon**, a list of tools will be displayed.
- Now, click on **Ribtop** in this list to activate the tool. On choosing this tool, the **Rib Top Feature** dialog box will be displayed along with the **SELECT SRFS** Menu Manager; refer to Figure-57. Also, you will be prompted to select floor of the reference model where you want to create the rib top feature. Note that rib feature is generally created after pocket milling or slot milling.
- Figure-58 shows an nc model with the material left on rib after pocket milling. Highlighted face is selected for rib top milling.
- Figure-59 shows the preview of the rib top feature to be created. Click on **OK** button from the dialog box to create the feature, the options of the dialog box have already been discussed.

Figure-57. Rib Top Feature dialog box and SELECT SRFS Menu Manager

Figure-58. NC Model with the face to be selected for ribtop feature

Figure-59. Preview of the ribtop feature

Bosstop

The **Bosstop** tool is used to machine the top face of a boss. The procedure to use this tool is discussed next.

• To access this tool, click on the down arrow displayed adjacent to **Ribtop** in the **NC Features & Machining** panel of the **Ribbon**, a list of tools will be displayed.

- Now, click on **Bosstop** in this list to activate the tool. On choosing this tool, the **Boss Top Feature** dialog box will be displayed along with the **SELECT SRFS** Menu Manager; refer to Figure-60. Also, you will be prompted to select floor of the reference model where you want to create the boss top feature. Note that boss top feature is generally created after step milling or slot milling.
- Figure-61 shows an nc model with the material left on top face of a boss after step milling. Highlighted face is selected for boss top milling.
- Figure-62 shows the preview of the boss top feature to be created. Choose **OK** button from the dialog box to create the feature, the options in the dialog box have already been discussed.

Figure-60. Boss Top Feature dialog box and SELECT SRFS Menu Manager

Figure-61. Face of the boss to be selected for the boss feature

Figure-62. Preview of the boss top feature

Free Form

The **Free Form** tool is used to machine at a specified path. The procedure to use this tool is discussed next.

- Click on the **Freeform** tool from the **NC Features & Machining** panel in the **Machining** tab of the **Ribbon**. On choosing this tool, the **Freeform Milling** dialog box will be displayed; refer to Figure-63.
- Choose desired tool from the **Cutting Tool** drop-down by clicking in the down arrow adjacent to **None** option. Now, we need to specify the custom path on which the tool will move.
- To create the tool path, select [icon] button adjacent to **Sketch** in the dialog box. On doing so, **SETUP SK PLN** Menu Manager (short form of Setup sketching plane) will be displayed, refer to Figure-64.

Figure-63. Freeform Milling dialog box

Figure-64. SETUP SK PLN Menu Manager

• Select the face or plane on which you want to create the tool path, refer to Figure-65. You will be prompted to specify the orientation plane by using the options in the **SKET VIEW** Menu Manager, refer to Figure-66.

Figure-65. Face selected for creating free form milling

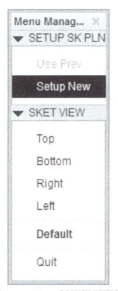

Figure-66. = SKET VIEW
Menu Manager

• Click on the **Default** option from the Menu Manager; the Sketcher environment will become active and the **References** dialog box will be displayed, prompting you to select three references for fully constraining the sketch; refer to Figure-67.

- Select two planes which are perpendicular to each other and a corner point, then choose the **Solve** button from the **References** dialog box and choose the **Close** button.

Figure-67. References dialog box

- Now, create the sketch of the tool path, refer to Figure-68 and click on the **OK** button from the **Ribbon**.

Figure-68. Sketch creating for tool path

- **Freeform Milling** dialog box will become active again. Specify the depth of cut by clicking on the Arrow button adjacent to **Cut depth** in the dialog box.
- On clicking this button, the **HEIGHT** Menu Manager is displayed; refer to Figure-69. By default, the Specify Plane option is selected in the Menu Manager. So, you can select a plane up to which you want to create the cut.

You can also specify the depth value from the plane on which you sketched the tool path. To do so, select the **Z Depth** option from the Menu Manager and specify the depth value in the message box displayed in the middle of the screen. After specifying the value choose the **Done/Return** option from the Menu Manager displayed and choose the **OK** button from the dialog box to create the tool path.

*Figure-69. HEIGHT
Menu Manager*

 Note
*Specify the value of depth of cut in negative otherwise the tool will move in positive Z direction
i.e. above the workpiece.*

The options in the Freeform Milling dialog box will be discussed in next chapter.

O-Ring

The **O-Ring** tool is used to machine grooves in the top face of a
workpiece. The procedure to use this tool is discussed next.

- Click on the **O-Ring** tool from the **NC Features & Machining** panel of
  the **Machining** tab in the **Ribbon**. The **O-Ring Feature** dialog box will
  be displayed along with the **SELECT SRFS** Menu Manager; refer to
  Figure-70. You will be prompted to select floor of the reference
  model where you want to create the O-ring feature. Select the face
  for the O-ring feature; refer to Figure-71 and click the **Done/Return**
  button from the Menu Manager.
- Figure-72 shows the preview of the O-ring feature displayed on
  choosing the **Done/Return** button.
- Click **OK** button from the dialog box to create the feature. The
  options in the dialog box have already been discussed.

Figure-70. O-Ring Feature dialog box and SELECT SRFS Menu Manager

Face to be selected

Figure-71. Face to be selected for O-ring feature

Figure-72. Preview of the O-ring feature

Undercut

The **Undercut** tool is used to machine undercuts in the workpiece. The procedure to use this tool is discussed next.

- Click on the **Undercut** tool from the **NC Features & Machining** panel of the **Machining** tab in the **Ribbon**. The **Undercut** dialog box will be displayed along with the **SELECT SRFS** Menu Manager, refer to Figure-73.
- You are prompted to select ceiling of the reference model where you want to create the undercut feature.
- Select the face for the undercut feature; refer to Figure-74 and click the **Done/Return** button from the Menu Manager. The preview of undercut feature will be displayed with all hard walls; refer to Figure-75.

Figure-73. Undercut dialog box and SELECT SRFS Menu Manager

Ceiling face
selected for
undercut feature

Figure-74. Face to be selected for creating undercut

Figure-75. Preview of the undercut feature with all hard walls

- We need to make outer walls as soft walls to create the feature. Click on the **Adjust Soft Walls** button from the **Undercut** dialog box.
- You will be prompted to select the walls to be shifted to soft walls. Select the outer two walls and click the **Done/Return** button from the Menu Manager displayed. The preview of the undercut feature will be displayed; refer to Figure-76.
- Click **OK** button from the dialog box to create the feature. The options in the dialog box have already been discussed.

Walls in cyan color are added to the soft walls

Figure-76. Preview of the undercut feature with the walls added recently in soft walls

Entry Hole

The **Entry Hole** tool is used to machine an entry hole for closed features like pocket feature, through pocket feature, O-ring feature and so on. The procedure to use this tool is discussed next.

- Click on the **Entry Hole** tool from the **NC Features & Machining** panel of the **Machining** tab in the **Ribbon**. The **Entry Hole** dialog box will be displayed; refer to Figure-77.
- By default, a closed nc feature is selected in the **Feature Name** drop-down of the dialog box.
- Specify the hole diameter in the **Hole Diameter** edit box and press **ENTER**.
- Select the reference model from drawing area. The **OK** button will be activated.
- Click on the **OK** button from the dialog box, the entry hole will be created.

Figure-77. Entry Hole dialog box

Hole Group

The **Hole Group** tool is used to machine a group of drill holes in the workpiece. The procedure to use this tool is discussed next.

- To create an hole group, choose the **Hole Group** tool from the **NC Features & Machining** panel of the **Machining** tab in the **Ribbon**. The **Drill Group** dialog box is displayed; refer to Figure-78.
- By default, **Axes** tab is selected in the dialog box. So, we need to select the axes of holes that we want to create.
- Axes can be selected in two ways; either one by one using the **Single** radio button in the dialog box or by using the pattern of the holes by selecting the **Pattern** radio button. Select the desired radio button.
- In our case, **Pattern** radio button is used. Select the **Pattern** radio button and then click on the **Add** button from the dialog box. You will be asked to select a hole.
- Select one of the holes in the pattern; refer to Figure-79 and press the middle button of mouse. The axes of all the holes will be displayed in the **Axes** current selected to drill area; refer to Figure-80.

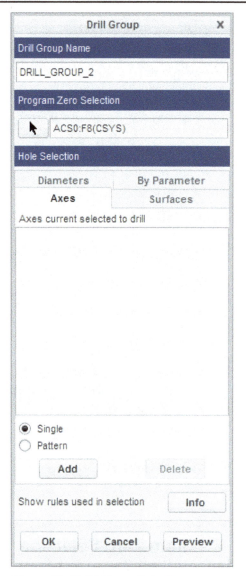

Figure-78. Drill Group dialog box

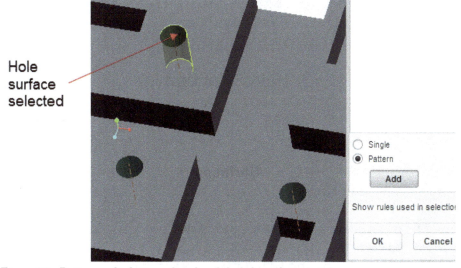

Figure-79. Pattern radio button selected and the hole surface selected

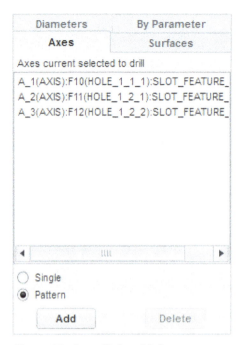

Figure-80. Axes of holes added

- Click on the **OK** button from the **Drill Group** dialog box to create the drill group feature. The options in the **Drill Group** dialog box are discussed next.

Drill Group Name

The **Drill Group Name** edit box is used to specify a user defined name for the drill group.

Program Zero Selection

The **Program Zero Selection** option is a kind of data collector. Click on the arrow adjacent to the box of this data collector. The system prompts to select a coordinate system. Select the desired coordinate system; the selected coordinate system will be used as program zero and its name will be displayed in the box of the collector.

Hole Selection

It is an area having four tabs; Axes, Surfaces, Diameters, and By Parameters. The options in **Axes** tab have been discussed while creating this feature. Rest of the options are discussed next.

Surfaces

The options in the **Surfaces** tab work in the same way as in **Axes** tab.

- Click on this tab and choose the **Add** button from the dialog box. You will be asked to select surfaces of the holes.
- Select a surface of the hole. If you want to add multiple holes then press and hold the **CTRL** key and click on the surfaces of the holes.
- Press the middle click of mouse and the selected surfaces will be added in the **Surface(s) for gathering holes to drill** area of the dialog box.

Diameters

The options in the **Diameters** tab are used to select the holes for the drill group feature according to their diameters.

- To add the diameters, select the **Diameters** tab and then click the **Add** button from the dialog box. The **Select hole diameter** dialog box is displayed; refer to Figure-81.

Figure-81. Select hole diameter dialog box

- Select a diameter from the dialog box and then click on the **OK** button from the dialog box. The holes of the selected diameter will be automatically added in the list and the diameter will be displayed in the **Diameter(s) of holes selected to drill** area; refer to Figure-82.

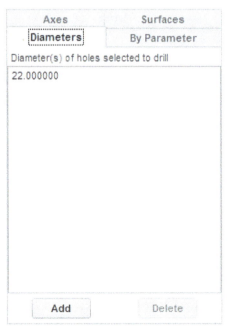

Figure-82. Diameter of selected holes

By Parameters

The options in the **By Parameters** tab are displayed as shown in Figure-83.

- Select the desired parameter from the **Feature Parameter** list box and select the value from the **Value** drop-down displayed below the list.
- By default, the = operator is selected in the **Operator** drop-down. For example, select the **DRILL DEPTH** parameter from the list; the

values of the holes available in the modeling area are displayed in the **Value** drop-down.

• Select one of the value from the drop-down and choose the **Add** button; all the holes satisfying the condition will be added in the drill group will be displayed at the bottom.

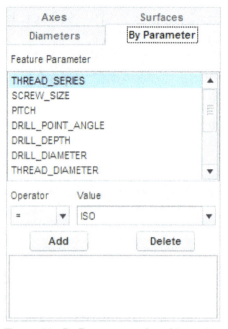

Figure-83. By Parameters tab and its options

Info

The **Info** tool is available at the bottom of the dialog box and is used to summarize about the holes selected for the drill group feature. On choosing this button, the information window is displayed as shown in Figure-84.

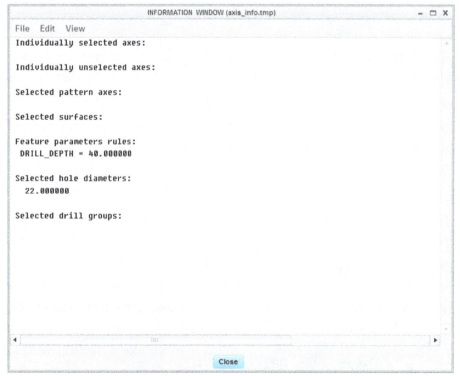

Figure-84. INFORMATION WINDOW

PRACTICAL 1

In this practical, you will create the nc features for the model displayed in Figure-85. From back side, the model is displayed as shown in Figure-86.

Figure-85. Model for Tutorial 1

Figure-86. Back side of the model

Starting Creo Expert Machinist

• Start Creo Parametric and choose the **New** button from the **Ribbon**.
• Select **Manufacturing** radio button from the **Type** area and **Expert Machinist** from the **Sub-type** area of the dialog box.
• Specify the name of the file as **C05 tut1** and clear the **Use default template** check box.

• Now, click the **OK** button from the dialog box, the **New File Options** dialog box will be displayed. Select the mmns mfg nc template and then choose the **OK** button.

Adding an NC model in the system

• Click on the **Create Model** button from **NC Model** panel in the **Ribbon**, a prompt will be displayed to define the name of the nc model.
• Click on the **Done** button ✅ from the prompt, the default name will be assigned and you will be prompted to select a part file as reference model.
• Select the file c5 model1 by using the options in the dialog box and then choose the **Open** button.
• Click on the **Create Stock** option from the **Menu Manager** displayed in the right. The **Auto Workpiece Creation** tab will be added in the **Ribbon** and the related options will be displayed.

Creating stock material(Workpiece)

• **Click on** the **Create Stock** button from **NC Model** Menu Manager displayed in right, drag handles and various values are displayed from the workpiece.
• Now, we will add 3 mm material on the top face and 3 mm material on bottom face of the workpiece. For that, enter 3 in both the **+Y** and **-Y** edit boxes displayed on clicking the **Options** tab; refer to Figure-87.

Figure-87. The Options tab with the values to be specified

• Click on the **Done** button ✅ from the **Ribbon** and select the **Done** button from the Menu Manager. The assembly environment will be displayed.

- Select the **Default** option from the **Constraint** drop-down and choose the **Done** button from the **Ribbon**; refer to Figure-88.

Figure-88. Default constraint in the Constraint drop-down

Creating Machine Setup

- After specifying the above parameters, the workpiece and the reference model will be placed at the default position i.e. coincident with the assembly coordinate system. Click on the down arrow displayed below **Work Center** in the **Setup** panel of the **Machining** tab in the **Ribbon**. A drop-down will be displayed.
- Choose the **Machine Tool Manager** tool from the drop-down; the **Milling Work Center** dialog box will be displayed.
- Click on the **Tools** tab at the top in the dialog box. The **Milling Work Center** dialog box will be displayed as shown in Figure-89.

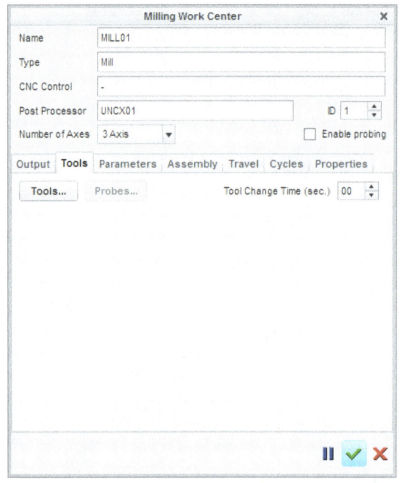

Figure-89. The Milling Work Center dialog box

- Click on the **Tools** tool from the dialog box. The **Tool Setup** dialog box will be displayed.
- Create the tools with the parameters as shown in Figure-90 and Figure-91.

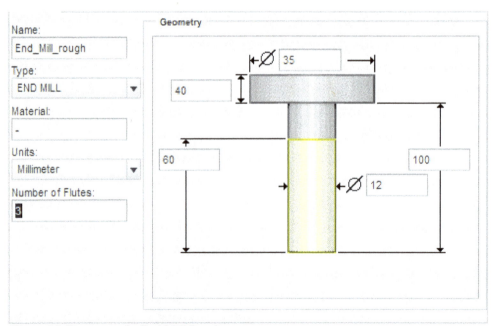

Figure-90. Parameters for End Mill Rough

Figure-91. Parameters for End Mill Finish

- Make sure that **Tool Number** of the End Mill Finish tool is set to **2** in the **Settings** tab of the **Tool Setup** dialog box. Now, choose the **OK** button from the **Tool Setup** dialog box and **OK** button from the **Milling Work Center** dialog box.

Creating an Operation

- Click on **Operation** tool from **Setup** panel of **Machining** tab in the **Ribbon**. On doing so, **Operation** tab open in the **Ribbon**.
- Click on the drop-down displaying **No Work Center** (i.e. **Work Center** drop-down). A list of machines will be displayed.
- Select the machine earlier created from the list i.e. MILL01. Now you need to select a coordinate system to specify the zero position of the machine.
- Click on the **Datum** button at the extreme right in the **Ribbon**. A drop-down will be displayed; refer to Figure-92.

Figure-92. Drop-down displayed on clicking the Datum button

- Click on **Coordinate System** tool in the drop-down, the **Coordinate System** dialog box will be displayed; refer to Figure-93.

Figure-93. NC Model with coordinate system and the Coordinate System dialog box

- Select the edges in such a way that the coordinate system is displayed as shown in Figure-93(in Isometric view).
- Click on **OK** button from the dialog box, the coordinate system is created.
- Click on Resume button ▶ from the **Ribbon**, now you are allowed to select the coordinate system.
- Select the coordinate system created recently and Click on the **Done** button ✔ from the **Ribbon**, an operation will be created and added in the Model Tree at the left.

Creating NC Features to remove material

Profile Feature

- Click on **Profile** tool from the **NC Features & Machining** panel, the **Profile Feature** dialog box will be displayed along with the **SELECT SRFS** Menu Manager.
- Select the side walls of the reference model while holding the *CTRL* key, refer to Figure-94.

Figure-94. Side walls to be selected

- After selecting the highlighted walls, Click on **Done** from the **SURF/ LOOP** Menu Manager and then Click on the **Done/Return** button from the **SELECT SRFS** Menu Manager, preview of the feature will be displayed.
- Click on the **OK** button from the **Profile Feature** dialog box to create the feature. The feature will be created and added below OP010 [MILL01] as PROFILE1 [OP010].

Slot Feature

- Click on the **Slot** tool from the **NC Features & Machining** panel in the **Ribbon**, the **Slot Feature** dialog box will be displayed along with the **SELECT SRFS** Menu Manager.
- Select the top face of the reference model while holding the CTRL key, refer to Figure-95.

Faces to be
selected

Figure-95. Faces to be selected for step milling

- After selecting the faces, press the middle button of mouse twice, the preview of the feature will be displayed.
- Choose **OK** button from **Slot Feature** dialog box, the feature will be created.

Creating Operation to machine from other side

Now, we need to machine the part from other side. For that we need to place the part in reverse manner i.e. upside-down on the machine. Same we need to do in Expert Machinist. This can be done by creating a new operation sequence. In this step, we will create a new Operation sequence.

- Rotate the part in such a way that the bottom face is displayed, refer to Figure-96.

Figure-96. Reverse side of the reference part

- Click on the **Operation** tool from the **Setup** panel in the **Machining** tab of the **Ribbon**, the **Operation** tab will open in the **Ribbon**.
- Select **MILL01** from the **Work Center** drop-down in the **Ribbon**.
- Click on the **Datum** button at the extreme right of the **Ribbon**, a drop-down is displayed.
- Select the **Coordinate System** tool ⚒ from the drop-down, the **Coordinate System** dialog box will be displayed.
- Select the edges in the such a way that the coordinate system is placed as shown in Figure-97.

Figure-97. The coordinate system to be created with the Coordinate System dialog box

- Choose the **OK** button from the dialog box and **Resume** button ▶ from the **Ribbon**.
- Now, select the recently created coordinate system and choose the **Done** button ✅ from the **Ribbon**. A new operation OP020 [MILL01] will be added in the **Model Tree**.

Creating NC Features to remove material

Pocket Feature

- Click on **Pocket** tool from the **NC Features & Machining** tab of the **Ribbon**, the **Pocket Feature** dialog box will be displayed along with the **SELECT SRFS** Menu Manager.
- Select the face of the reference model as shown in Figure-98 to remove material in the form of a pocket.

Figure-98. Face to be selected to create a pocket feature

- Press the middle button of mouse twice, preview of the pocket feature will be displayed.
- Click on **OK** button from the dialog box to create the feature.

Similarly, create the other pocket features one by one by selecting the corresponding faces. The model should be displayed as Figure-99 after choosing the **Material Removal Display** button from the **Display** panel of the **Machining** tab. Now, the only material left on the reference model is at the faces of the ribs. Here, we will remove this material by using the Slab feature, because it removes the material from a large area and with same tolerance.

Slab Feature

- Click on **Slab** tool from the drop-down displayed on clicking at the down arrow adjacent to **Flange** in the **NC Features & Machining** tab of the **Ribbon**, the **Slab Feature** dialog box will be displayed along with

the **SELECT SRFS** Menu Manager.
- Select the face of the model as shown in Figure-99.

Figure-99. Face to be selected for the slab feature

- Press the middle button of mouse twice, preview of the slab feature will be displayed as shown in Figure-100.

Figure-100. Preview of the slab feature

- Click on the **OK** button from the dialog box to create the feature.

Hole Group

- Click on **Hole Group** tool from the **NC Features & Machining** tab of the **Ribbon**, the **Drill Group** dialog box will be displayed.
- Click on **Add** button from the dialog box, you are allowed to selected the faces to add their corresponding axes in the list.
- Press and Hold the **CTRL** key and select faces of the holes, refer to Figure-101.

Faces to be selected

Figure-101. Faces to be selected for creating holes

- Press the middle mouse button, axes will be added in the **Axes current selected to drill** list box.
- Click on the **OK** button from the dialog box, the feature will be created and the material would be removed.

Press **CTRL+S** to save the nc model.

PRACTICAL 2

In this tutorial, you will create the nc feature for the model in Figure-102. You can request the resources kit of this book at cadcamcaeworks@gmail.com

Figure-102. Model for tutorial 2

Starting Creo Expert Machinist

- Start Creo Parametric and choose the **New** button from the **Ribbon**.
- Select **Manufacturing** radio button from the **Type** area and **Expert Machinist** from the **Sub-type** area of the dialog box.
- Specify the name of the file as **C05 TUT2** and clear the **Use default template** check box.
- Now, click on the **OK** button from the dialog box, the **New File Options** dialog box will be displayed. Select the **mmns mfg nc** template and then click on the **OK** button.

Adding an NC model in the system

- Click on the **Create Model** button from **NC Model** panel in the **Ribbon**, a prompt will be displayed to define the name of the nc model.
- Click on the **Done** button from the prompt, the default name will be accepted and you will be prompted to select a part file as reference model.
- Select the file C05 TUT2 by using the options in the dialog box and then click on the **Open** button.
- Click on the **Create Stock** option from the **Menu Manager** displayed in the right side of the screen. The **Auto Workpiece Creation** tab will be added in the **Ribbon** and the related options will be displayed.

Creating stock material

- Click on the **Create Stock** button from **NC Model** Menu Manager displayed in right, drag handles and various values are displayed from the workpiece.
- Now, we will add 3mm material on the top face and all other faces except bottom face of the workpiece. For that, enter 3 in +Y,+X,-X,+Z, and -Z in respective edit box from **Options** tab; refer to Figure-103.

Figure-103. The Options tab on specifing the values

- Click the **Done** button from the **Ribbon** and click the Done button from the menu manager, assembly environment will be displayed.
- Select the **Default** option from the **Constraint** drop-down and choose the **Done** button from the **Ribbon**.

Creating Machine Setup

- After specifying the various parameters, the workpiece and the reference model will be placed at the default position i.e.coincident with the assembly coordinate system. Click on the down arrow displayed below **Work Center** in the **Setup** panel of the **Machining** tab in the **Ribbon**, a drop-down will be displayed. Click on the **Machine Tool Manager** tool from the drop-down. The **Milling Work Center** dialog box will be displayed.
- Click on the tools tab at the top in the dialog box, the Milling Work box will be displayed as shown in Figure-104.

Figure-104. The Milling Work Center dialog box

- Click on the tools from the dialog box, the **Tool Setup** dialog box will be displayed.
- Create the tools with parameters as shown in Figure-105 and Figure-106.

Figure-105. Parameters for End Mill Rough

Figure-106. Parameters of End Mill Finish

- Make sure that **Tool Number** of the End Mill Finish tool is set to 2 in the **Settings** tab of the **Tool Setup** dialog box. Now, choose the **OK** button from the **Tool Setup** dialog box and **OK** button from the **Milling Work Center** dialog box.

Creating an Operation

- Click on **Operation** tool from **Setup** panel of **Machining** tab in the **Ribbon**. On doing so, Operation tab open in the Ribbon.
- Click on the drop-down displaying **No Work Center** (i.e. **Work Center** drop-down), a list of machines will be displayed.
- Select the machine earlier created from the list i.e. MILL01. Now you need to select a coordinate system to specify the zero position of the machine.
- Click on the **Datum** button at the extreme right in the **Ribbon**, a drop-down will be displayed; refer to Figure-107.

Figure-107. Drop-down displayed on clicking the Datum button

- Click on the coordinate tool in the drop-down, the **Coordinate System** dialog box will be displayed; Figure-108.

Figure-108. NC Model with the Coordinate System dialog box

- Select the Top face of the Stock and adjust the coordinate system as displayed in above figure.
- Click on the **OK** button from the dialog box. The coordinate system is created.
- Click on Resume button ▶ from the **Ribbon**, now you are allowed to select the coordinate system.
- Select the coordinate system created recently and Click on the **Done** button ✓ from the **Ribbon**, an operation will be created and added in the Model Tree at the left.

Creating NC Features to remove material

Profile Feature

- Click on **Profile** tool from the **NC Features & Machining** panel, the **Profile Feature** dialog box will be displayed along with the **SELECT**

SRFS Menu Manager.
- Select the side walls of the reference model while holding the **CTRL** key; refer to Figure-109.

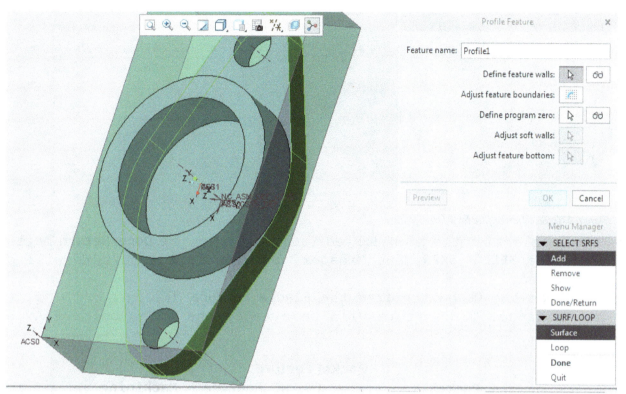

Figure-109. Side walls to be selected for Profile option

- After selecting the highlighted walls, Click on **Done** from the **SURF/ LOOP** Menu Manager and then Click on the **Done/Return** button from the **SELECT SRFS** Menu Manager, preview of the feature will be displayed.
- Click on the **OK** button from the **Profile Feature** dialog box to create the feature. The feature will be created and added below OP010 [MILL01] as PROFILE1 [OP010].

Flange Feature

- Click on the **Flange** tool from the **NC Features & Machining** panel in the **Ribbon**, the **Flange Feature** dialog box will be displayed along with the **SELECT SRFS** Menu Manager.
- Select the face of the reference model as shown in Figure-110.

Figure-110. Selection of face for Flange feature

- After selecting the highlighted face, Click on the **Done/Return** button from the **SELECT SRFS** Menu Manager, preview of the feature will be displayed.
- Click on the **OK** button from the **Flange Feature** dialog box to create the feature. The feature will be created and added below FLANGE1 (OPO10).

Pocket Feature

- Click on the **Pocket** tool from the **NC Feature & Machining** tab of the **Ribbon**, the **Pocket Feature** dialog box will be displayed along with the **SELECT SRFS** Menu Manager.
- Select the face of the reference model as shown in Figure-111 to remove material in the form of a pocket.

Figure-111. Selection of face for pocket feature

- Press the middle button of mouse twice, preview of the pocket feature will be displayed.
- Click on the OK button from the dialog box to create the feature.
- Click on the **Material Removal Display** button from **Ribbon** to see the current preview of the specific feature.

Hole Group

- Click on **Hole Group** tool from the **NC Features & Machining** tab of the **Ribbon**, the **Drill Group** dialog box will be displayed.
- Click on **Add** button from the dialog box, you are allowed to selected the faces to add their corresponding axes in the list.
- Press and Hold the **CTRL** key and select faces of the holes, refer to Figure-112.

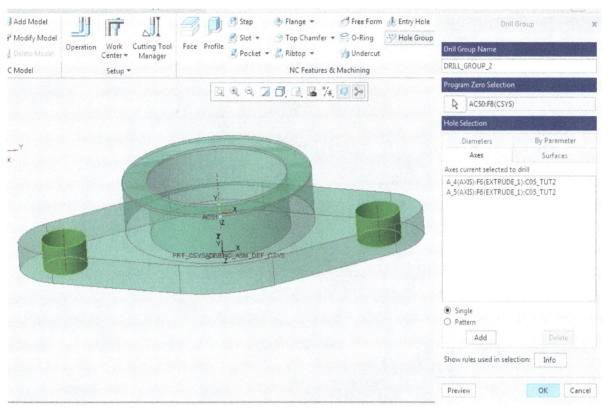

Figure-112. Selection for Hole Group feature

- Press the middle mouse button, axes will be added in the **Axes current selected to drill** list box.
- Click on the **OK** button from the dialog box, the feature will be created and the material would be removed.

Press **CTRL+S** to save the nc model.

PRACTICE 1

In this problem, you will import a reference model, as shown in Figure-113 & Figure-114 and you will create a workpiece having uniform thickness of 2 over the reference model. Then you will create a machine setup including the required tools for various operations on the workpiece. Then you will create all the required nc features. The model file for reference model is available in the Resources kit.

Figure-113. Isometric view of the model

Figure-114. Reverse side view of the model

PRACTICE 2

In this problem, you will import a reference model, as shown in Figure-115 & Figure-116. Then you will create a machine setup including the required tools for various operations on the workpiece. Then you will create all the required nc features. The model file for reference model is available in the Resources kit.

Figure-115. View for Problem 2

Figure-116. View for Problem2

FOR STUDENT NOTES

Chapter 6

Creating toolpaths and Output

Topics Covered

The major topics covered in this chapter are:

- *Generating Tool paths*
- *Various analysis that are performed on tool paths*
- *Checking for possible gouges*
- *Checking for tool clearances*
- *Mimicking a tool path*
- *Creating output file*

TOOL PATHS

Tool path is an imitation of the path at which tool moves while cutting the material from the workpiece. While doing a cutting operation, the tool does not move at random paths. But, it follows a specific path. In Creo Expert Machinist, there are various patterns (tool paths) that a tool follows while doing a specific operation. These patterns are applied by using the dialog boxes that are displayed while creating tool paths for various operations. For example, when you have created a face milling feature on the workpiece then you are required to create a tool path for that feature and while creating the tool path for face milling feature, the options related to face milling are displayed in the related dialog box. In the previous chapter, we have create face milling feature and other such nc features. Here, we will generate tool paths for these features one by one.

Tool path for a Milling

Select a milling feature like; **PROFILE1 [OP010]** from the **Model Tree** displayed at the right of the modeling area and right-click on it. A shortcut menu will be displayed; refer to Figure-1.

Figure-1. Shortcut menu for creating tool path

* Click on the **Create Toolpath** option from the shortcut menu, the **Face Milling** dialog box will be displayed as shown in Figure-2.

Note

You can also generate tool path of an nc feature by using **Machining** tool available at the right-most in the **NC Features & Machining** panel of the **Machining** tab in the **Ribbon**. On choosing this tool, **Select Feature** list box is displayed. Select a feature from the list box and choose the **OK** button available below this list box, the related tool path generation dialog box will be displayed.

Figure-2. Profile Milling dialog box

• Click on the down arrow displayed adjacent to the **Cutting Tool** in the dialog box. List of the tools created earlier will be displayed.

• Select an appropriate tool from the list and click on the **OK** button from the dialog box, the toolpath will be created with the name **feature name TPnumber** and will be listed below its nc feature in the Model Tree. Here, **feature name** is the name of feature for which you are creating the tool path. **number** is the sequence number of the toolpath for current feature.

Note

The list displayed on clicking the down arrow of Cutting Tools is sorted and filters automatically depending on the feature for which you are creating the tool path. For example, if you are creating tool path for creating undercut then the tools that are capable of under cutting will only be displayed in the list.

Figure-3 shows a tool path created and added below the Profile nc feature.

Figure-3. Toolpath created for Profile NC feature

Various options available in the **Profile Milling** dialog box and their effects are discussed next.

Tool Path Name

The **Tool Path Name** is an edit box used to specify the name of the tool path by which it will be listed in the **Model Tree** and its output will be generated. The default pattern of naming in this edit box is **feature name TPsequence number [Operation Number]**. You can specify any desired name in the edit box. But the maximum number of characters that can be specified are 32.

Comments

The **Comments** button is used to display a window in which you can write the comments related to the current cutting strategy(tool path). On choosing this button, the **Machine Strategy Comments** window is displayed as shown in Figure-4.

Figure-4. **Machine Strategy Comments** *dialog box*

- Write the comments related to the tool path in the text box of this window.
- You can open an already existing text file by using the Open button available in the **File** menu of this window.
- After writing the text you can save it as a text file by using the **Save** or **SaveAs** button available in the **File** menu.
- Also, you can insert matter written in a text file by using the **Insert** button available in the **Edit** menu.
- After specifying the comments, click on the **OK** button available at the bottom of the window to apply the comment.

Feature Name

The **Feature Name** box displays name of the feature for which you are creating the tool path. You can not edit the value displayed in this box.

Preview

The **Preview** button is used to display the preview of the nc feature (in the modeling area) for which you are creating the tool path. On choosing this button, the feature is highlighted in blue color in the modeling area.

Cutting Tool

This is an area having a button named **Access Tool Manager** and a drop-down list having list of tools available in the machine setup.

- On choosing the **Access Tool Manager** button, the **Tool Setup** dialog box will be displayed.
- This dialog box and its options have already been discussed in Chapter 3 (Tool Setting).
- You can either create new tool by using the options in the **Tool Setup** dialog box or you can select from the already existing tool available in the drop-down list.

Show Tool

The **Show Tool** button is used to display preview of the tool selected from the list. On choosing this tool, a preview window is displayed having name of the tool as the Title of the window; refer to Figure-5. You can zoom in/out and pan in this window by using the mouse gestures.

Figure-5. Preview window

Machining Mode

There are four section available in this area. These options are used to specify the machining process. The sections available in this area are; **Wall machining, Cut Motion, Motion between cuts, Options,** and **Start Position.** These sections are discussed next.

Wall machining

There are five options are available in this area which are discussed next.

- Select the **Rough** check box, if you want to do roughing operation of the workpiece. The **Rough to** edit box becomes available. Enter the amount of stock to remain on the hard walls after roughing in the edit box. This edit box is discussed next.
- Select the **Finish check box**, if you want to do the finishing process on the workpiece. This is the cutting pass in which accuracy of the machine is achieved.
- Click on the **Finish Cuts** button, if you want to specify the multiple cuts at multiple offsets from the hard walls. On clicking this button, the **Finish Cuts** dialog box will be displayed. This dialog box is used to specify the number of passes that will be required to do finishing on the workpiece.
- Select the **Corners only** check box to remove the material only from corners of a model. In our case this check box is disabled.
- Select the **Use CUTCOM** check box, if you want to use cutter diameter compensation for this tool path.

Cut Motion

There are many options in this section which are discussed next.

- Select the **One Direction** check box, if you want to cut the material in only one direction. Note that when the tool passes a cutting operation, the tool moves upward i.e. up to the clearance plane; moves to the starting line and then again does the next cutting pass.
- Check the **Back and Forth** check box, if you want to cut the material in both the motions of tool i.e. Forward motion as well as the reverting back of tool to its starting line.
- Select the **Climb** check box during one directional cut, if you want to cut the material using the climb cutting motion. The tool rotates in Counter Clockwise direction with respect to stock material.
- Select the **Conventional** check box during one directional cut, if you want to cut the material in the conventional manner i.e. the tool rotates clockwise with respect to the stock material.
- Select the **Follow Hard Walls** check box, if you want that each cut will follow the contour of the hard walls of the feature.
- Select the **Straight Line** check box, if you want to cut the model ain a straight cut.
- Click in the **Cut Angle** edit box and enter the specific value of cut.
- Check the **Clean Cut** dialog box, if you want to clean the scallops from the straight cut.
- Click in the **Stock** edit box and enter the value to clean up the cuts.

Motion Between cuts

There are four options in this section which are discussed next.

• Select the **Clear Cut** check box if you want to move the tool upwards in a non-cutting pass. Selecting this radio button makes the tool move away from the workpiece in upward direction so that it can return to its starting line in rapid speed.

• Select the **Stay Down** check box if you want the tool to remain at cut depth while connecting to the next cut.

• Select the **Stay In Cut** check box if you want the tool to remain engaged in the material at the end of each cut.

• Select the **Retract** check box if you want to remain the tool retract before connecting to the next cut.

Start Position

• Select the **Use Default** check box if you want to start the cutting from default position.

Tool Path Properties

This button is used to specify various properties related to the tool path. On choosing this tool, **Tool Path Properties** dialog box is displayed; refer to Figure-6. The options are discussed after the completion of the options of **Profile Milling** dialog box.

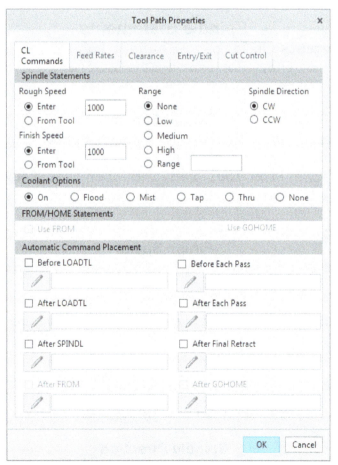

Figure-6. Tool Path Properties dialog box

Options

• Select the **Use Fixture Offset** check box, if you want to add offset to the original tool path because of a fixture installed on the machine. This fixture can be for the workpiece as well as for the tool.

- After specifying various parameters, click on the **OK** button from the dialog box to create the tool path and add it to its nc feature in the **Model Tree**.
- Select the newly created toolpath from the **Model Tree** and click on the **Tool Path Player** tool from the **Output** panel in the **Ribbon** to display the simulation of tool and workpiece while tool moves on the tool path being created. We will discuss more about this tool later in this chapter.

TOOL PATH PROPERTIES DIALOG BOX

The **Tool Path Properties** dialog box is displayed on choosing the **Tool Path Properties** button from the Face Milling dialog box. There are five tabs in this dialog box; **CL Commands**, **Feed Rates**, **Clearance**, **Entry/Exit**, and **Cut Control**. These tabs and their options are discussed next.

CL Commands

The **CL Commands** options in this tab are used to modify the entries related to CL commands like, spindle speed, rotation direction, and so on. The options in this tab and their effects are discussed next.

Spindle Statements

The **Spindle Statements** section is used to set values of variables related to spindle like, direction of rotation, range, roughing speed and so on.

Rough Speed

- Select the **Enter** radio button for **Rough Speed** to specify the value in RPM for rough cutting speed.
- If you select the **From Tool** radio button for **Rough Speed**, then the cutting speed will be assumed as the speed specified for the current tool.

Finish Speed

The options for **Finish Speed** work in the same way as for **Rough Speed**.

Range

There are five radio buttons available to specify the spindle RPM range in which the spindle speed can vary.

- You can specify a used defined value by selecting the **Range** radio button. An edit box adjacent to the radio button becomes active where you can specify the value of range.

Spindle Direction

There are two radio buttons to specify direction in which the spindle will rotate.

- You can rotate the spindle either clockwise direction **(CW)** or in counter clockwise direction **(CCW)** by selecting the respective radio button.

Coolant Options

There are six radio buttons available in this section which are discussed next.

- Select the **ON** radio button if you want the coolant to flow in the normal mode i.e. through the pipes aimed at the workpiece directly.
- Select the **Flood** radio button if you want the coolant flow at high speed and through all the vents made for coolant in the working area.
- Select the **Mist** radio button if you want to create the mist of coolant in working area.
- Select the **Tap** radio button if you want the coolant to flows through the tool vents. This radio button is selected while doing a tapping cycle.
- Select the **Thru** radio button if you want the coolant to pass through the tool as well as through the coolant pipes.
- Select the **None** radio button if you do not want the coolant to flow on the workpiece.

FROM/HOME Statements

There are two check boxes in this section which are discussed below.

- Select the **Use FROM** check box if you want to manually add some CL commands with the **FROM** statement in the output.
- Select the **Use GOHOME** check box if you want to manually add some CL commands with **GOHOME** statement.

On selected these check boxes, the respective check boxes in the **Automatic Command Placement** area become active.

Automatic Command Placement

The options in this section are used to customize the CL output automatically.

- Select any of the check box in this area and then you can specify the user-defined CL commands before or after the command corresponding to the selected check box.
- For example, if you want to stop the coolant before loading a tool then select the **Before LOADTL** check box and then specify the command in the corresponding edit box.
- If you do not remember the command then select the **Edit** button the **CL Command** dialog box will be displayed; refer to Figure-7.

Figure-7. CL Command dialog box

- Now, either you can use an already existing CL file by using the **File** button in the dialog box or you can use the menu containing all the CL codes. Click on the **Menu** button to display the **cmd edit** Menu Manager, refer to Figure-8.

Figure-8. cmd edit Menu Manager

- For example, click on the **COOLNT-FROM** option then **COOLNT** option and then **OFF** option from the Menu Manager. The command will be added in the **CL Command** dialog box. Click the **OK** button from the dialog box, the command will be added before **LOADTL** command in the CL output.

Feed Rates

The options in the **Feed Rates** tab are used to specify the feed rates for various operations being done on the workpiece like, positioning, retract, rough cutting, side entry, and so on. Some of the options will not be available based on the tool path selected. For example for Facing operation the Positioning options will not be available.

You can set the value of feed rate for operation by using the two methods; either by specifying the value in edit box or by using the data specified for the current tool. The method of setting feed rate for Roughing is given next. The method for setting feed rate for other will be the same.

* Select the unit in which you want to specify the feed rate by selecting corresponding radio button from the **Feed Units** section of the tab. In this case, we will use **MMPM** radio button.
* Click on the down arrow displayed below the **Rough Cutting**, a list of options will be displayed. Select **Enter** option if you want to enter a value manually or select **From Tool** option if you want to use the tool data specified for the current tool. In this case, we will select the **Enter** option.
* On selecting the **Enter** option the edit box adjacent to the drop-down will be active. Enter the desired value of feed in the edit box.

Clearance

The options in the **Clearance** tab are used to specify the clearance distance between the workpiece and the tool while doing specific cutting passes displayed in this tab. You can specify the clearance distance for following cutting passes in the dialog box.

* **Initial Approach**
* **Final Withdrawal**
* **Between Cuts**
* **Between Passes**

There are Five ways for specifying clearance distance for each of the above cutting pass. These ways are discussed next.

Along Z-Axis

Select the **Along Z-Axis** radio button if you want to specify the clearance distance for the selected cutting pass along the Z-axis. On selecting this radio button, the edit box adjacent to the selected radio button will become active. Now, enter the desired value in the edit box. The distance specified will be calculated from the Coordinate System selected for the Operation.

Clear Workpiece

Select the **Clear Workpiece** radio button if you want to specify the clearance from the outermost face layer of the workpiece. The clearance distance specified in the adjacent edit box will be calculated along the Z direction from the top face of the workpiece.

Clear Feature

Select the **Clear Feature** radio button, if you want to specify the clearance from the reference model. Note that in this case also the direction will be along the Z direction.

Go Delta

Select the **Go Delta** radio button, if you want to specify the distance value from the last cut tool path. The direction here will also be along the Z direction.

None

Select the **None** radio button, if you do not want to specify the clearance distance. Note that this radio button is not available for the clearance specified for **Between Cuts**. For Between Cuts, you must have to specify the clearance value.

Indepth

The values specified for the clearance play very important role in the machine output. The wrong values specified in these edit box can restrict the tool path creation by this software and you might get the error as **Check Setup** in the information bar displayed below in the application window. In some cases, it may cause an accident in the real time machining if the user ignores these value. So, this is recommended that you take extra care while specifying values in the edit boxes of the **Clearance** tab.

Entry/Exit

The options in the **Entry/Exit** tab are used to specify the options related to entry and exit of the tool for each cutting step. There are two areas in this dialog box; Setup Landing/Lift and Side clearance.

Setup Landing/Lift

There are two categories available in this section i.e. **Soft Landing** and **Soft Lift**.

- Select the desired radio buttons for the **Soft Landing** and **Soft Lift**.
- There are three radio buttons for each of the categories in the dialog box; **Automatic**, **Always**, and **Never**.
- If you select the **Automatic** radio button, then the soft lift and landing will automatically be applied on the tool path.
- Select the **Always** value if you want to specify the value for landing and lift distance manually for the tool path. On selecting this radio button the corresponding check box will become active automatically. The specified value will be according to the unit system selected.
- Select the **Never** radio button if you do not want to specify the landing/lift in the tool path.

Side clearance

There is one edit box available in this area. Specify the value of side clearance in the edit box.

Cut Control

The options in the **Cut Control** tab are used to specify parameters for the cut. The options in this are discussed next.

Maximum Cut

The options in this section are used to specify the depth of cut or step over limit for the cut. There are two options in this area; **Depth of Cut** and **Stepover**. You can specify the depth of cut for each cutting pass by using the **Depth of Cut** options. This is the vertical distance between one tool path and the successive tool path. The **Stepover** option is used to specify the horizontal distance between one cutting pass and the other consecutive cutting pass.

Indepth

The step over value is used at the same cutting plane and the depth of cut is the distance between such two cutting planes, refer to Figure-9.

Figure-9. Step over value and Depth of Cut value

CUTCOM Control

The options of **CUTCOM Control** section is active when **Use CutCom** check box is selected from Profile Milling/other cutting operation dialog box.

- Click in the **Register** edit box and enter the number of machine offset register to be used for tool path.
- Click on the **On/Off Move** edit box and enter the length of linear move during which **CUTCOM** will be invoked and cancel for this tool path.
- Select the **Perpendicular** radio button, if you want the linear move will be perpendicular to the following move of the tool path.
- Select the **Tangent** radio button, if you want the linear move will be tangent to the following move of the tool path.

Options

There are various options in this section which are discussed next.

- Click in the **Tolerance** edit box and enter the value for tolerance.

Stay in Cut

The options in this area are used to specify the extent up to which the tool will remain engaged with the material while cutting.

- Select the **On** option from the **Tool Center** drop-down if you want to move the tool away from the workpiece when the center of the tool is exactly on the edge of soft wall.
- Select the **To** option from the drop-down if you want to move the tool away from the edge when tool reaches at specified distance from the edge of soft wall. The distance value can be specified in the edit box available below this drop-down.
- If you select the **Past** option from the drop-down list then tool will cross the specified distance from the edge of soft wall and then will move away.

Circle Moves

- Select the **Points Only** radio button, if you do not want to use CL file output in circular interpolation. It uses the points only for output.
- Select the **Arc only** radio button if you want to use output circle statement and number of points for arc in CL data.
- Click in the **Number of Arc points** edit box and specify the value for the points of arc.

Corners condition

The options of this section are used for cutting in corners.

- Select the Fillet radio button, if you want to make corners round.
- Select the Straight radio button, if you want to make corners straight.

After specifying the various parameters, click **OK** button from **Tool Path Properties** dialog box.

TOOL PATH PLAYER

The **Tool Path Player** tool is available in the **Output** panel of **Ribbon**. This tool is used to simulate the cutting motion of tool on the tool path. This tool is active only when you select an operation having toolpaths or a tool path from the **Model Tree**. The procedure to use this tool is discussed next.

- Click on the **Tool Path Player** tool. The system will process all the information provided by you in various relevant dialog boxes and after the completion of process, the **PLAY PATH** dialog box will be displayed with the preview of tool in the Modeling area; refer to Figure-10.
- Choose the **Play Forward** button ▶ from the dialog box to play the simulation of the tool. The system will start simulation of tool path.

Figure-10. PLAY PATH dialog box with the preview of tool

The options available in this dialog box are discussed next.

File Menu

The options in the **File menu** are related to file handling i.e. opening file, saving file and so on. Using the **Save as MCD** option from this menu, you can save the CL(Cutter Location) file as MCD (Machine Control Data) file. A MCD file includes the data related to Post Processor and various checks that can be performed on the CL file.

View Menu

The options is **View menu** are more related to analysis of tool path. Various options available in this menu are discussed next.

Wireframe

The **Wireframe** option is selected by default. As a result, the tool path is displayed in the form of wireframe i.e. in red colored lines in this case. This option is inactive in the menu as you can not deselect this option until you select any alternate option.

NC Check

The **NC Check** option is selected when the tool path is to be simulated on a machine. Click on the **NC Check** option from the View menu.

Show Remaining Material

The **Show Remaining Material** option is used to perform material analysis. In some cases, a specific type of tool is not able to cut material from the workpiece in certain circumstances. This option generally available while doing turning process so for Expert Machinist it would not be generally available.

Display Tool Tip

The **Display Tool Tip** option is selected when you want to display the trajectory of the tool tip while removing the material for the current operation.

Display Tool Kerf

The **Display Tool Kerf** option is selected when you want to display the impression of perimeter of tool while removing material from the workpiece.

Shaded Tool

The **Shaded Tool** option is selected when you want to display the tool in shaded style. If you deselect this option, the tool will be displayed as wireframe model.

Collision Checking

The **Collision Checking** option is available for the complete assembly of machine and fixtures. If you have selected this option then the volume that is interfering with the machine or fixture setup will be displayed in red color. Note that the collision between tool and workpiece can not be displayed using this option.

Stop at Collision

The **Stop at Collision** option if selected will force the tool to stop while simulating tool path as soon as the tool collides with interfering objects.

NCL File Menu

The options in the **NCL File Menu** menu are used to operate and edit the Numeric Cutter Location (NCL) file. Using the options in this menu, you can control the way the tool will simulate by following the CL file. You can insert a user define command in the current CL file and check its simulation.

CL Data

▶ CL Data   This node is selected when you want to check the CL file while simulating the tool path. To display the CL data, click on the **CL Data** node displayed below the menu bar. On selecting this node, the node expands and the CL Data file is displayed; refer to Figure-11.

Figure-11. PLAY PATH dialog box with expanded
CL Data node

Below the CL Data area, there are various buttons to perform the operations like starting, stopping, forwarding and so on. These buttons are the standard options to play simulations. The slider displayed below the buttons is used to increase or decrease the speed of simulation. Other buttons and options are discussed next.

Tool Clearance

The **Tool Clearance** button is used to check the tool clearance while the tool follows the tool path. On selecting this button, the **MEASURE** Menu Manager is displayed; refer to Figure-12. The options in this menu manager are similar to the tools available in the **Measure** drop-down in the **Measure** panel of the **Analysis** tab of the **Ribbon**. The result of measurement is displayed in the Information Bar available below the modeling area.

Gouge Check

The **Gouge Check** button is used to check the clearance and other parameters of tool path by using the gouge.

• Click on the **Gouge Check** button. The **MFG CHECK** Menu Manager is displayed.

- Select the surface for which you want to perform gouge check. Press the middle mouse button three times to exit the selection process.
- On doing so, the **GOUGE CHECK** Menu Manager will be displayed as shown in Figure-13.
- Select the desired options to specify the related parameters for the gouge check and then click on the **Run** option from the Menu Manager to perform the check. Make sure that you have already setup a probe for gouge check.

Figure-12. MEASURE Menu Manager

Figure-13. MFG CHECK Menu Manager

Position Cutting Tool

This option is available at the right of the **Gouge Check** button. By selecting this option, you can position the tool at any specific point on the tool path. The procedure to use this tool is discussed next.

- Click on this option and then click on the position where you want to place the tool. On doing so, the tool will be placed at the specified point and the cutting step corresponding to the selected point will be highlighted in the CL Data.

Till this point, we have simulated the tool path and checked for any error in the tool path. If the tool path is OK then we are ready to generate the output file of the tool path so that the machine can work on it. The next topic is about generating the output files.

GENERATING OUTPUT FILES

There are two options in Creo Expert Machinist to generate the output files. You can generate output as cutter location data or you can create a post processed output file directly feed able to the machine. Both the methods are discussed next.

Generating Cutter Location Data

To generate cutter location data, select the toolpath from the **Model Tree** and then select **Create CL File** tool from **Output** panel in the **Machining** tab of the **Ribbon**. On doing so, the **Save a Copy** dialog box will be displayed; refer to Figure-14.

Figure-14. Save a Copy dialog box

• Specify the name of the file in the dialog box and then click **OK** button from the dialog box to save the file. The file will be saved with an extension **.ncl**. You can preview the file in any word processor. Figure-15 shows a cutter location data file in WordPad application of Microsoft Windows.

Figure-15. A cutter location data file opened in Microsoft WordPad

Here, you can edit the file and then using any nc post processer application, you can generate the output file. The files that machine read are of **.tap** extension.

Generating post processed output file

 ▾ Using this method you can directly create the output file for the machine. The procedure to use this tool is discussed next.

- Click on the down arrow adjacent to **G00** button in the **Output** panel of the **Machining** tab in the **Ribbon**.
- There are two tools available in this list i.e. **Automatic** and **Select Post**; refer to Figure-16.

Figure-16. Tools for output

- On selecting the **Automatic** tool, the **Save a Copy** dialog box will be displayed and you are asked to save the cutter location file first. Click on the **OK** button from the dialog box and the output file with **.tap** extension will be saved in the same directory. Note that the post processor used will be the one specified while setting the machine in Machine Setup. This has been discussed earlier in chapter 4.
- On selecting the **Select Post** tool from the list, the **Save a Copy** dialog box is displayed and you are asked to save the CL file first. Save the file by choosing the **OK** button. As soon as you save the file, **POST PROCESSOR LIST** (**PP LIST**) Menu Manager is displayed, refer to Figure-17.

Figure-17. PP LIST Menu Manager

- Select a post processor of your choice from the list, an **INFORMATION WINDOW** with the information related to the output will be displayed; refer to Figure-18.

Figure-18. The INFORMATION WINDOW

- Click on the **Close** button after reading the information and then open the file saves with **.tap** extension in a word processor to see the output of the tool path. Figure-19 show the output of a tool path opened in WordPad application.

Figure-19. The output file with .tap extension opened in WordPad

Now, you can edit the file and then feed it to the machine to get the machining output.

Till this point, we have covered the generation of output after specifying all the parameters whether it is related to machine, workpiece, tool geometry, operation setting, or tool path. This is the point where I want the users of this book, to go back to previous chapters and then check the final output of same operation by using different settings applied in Machine Setup and Tool Setup. This will give an idea about, how the settings affect on output. For example, if we change the origin of operation; will the coordinates in output file be the same?

You have used a post processor available by default in Expert Machinist. Now, is it sure that it will work for your machine. In many cases, the answer will be **No**. So, we need to create a post processor which is suitable for your machine.

Creating a post processor is a cumbersome job but we can modify a post processor in such a way that it understand the requirements of your machine. This is done by using the Option file. An Option file contains the data related to the machine which is used by Post Processor for generating output. To create an Option file for Expert Machinist, there is a separate App in Creo called **NC Post Processor**. To access this App, click on the **Applications** tab in the **Ribbon**; refer to Figure-20. All the manufacturing applications will be displayed in the **Ribbon**. Select the **NC Post Processor** button, a separate application will open named **Option File Generator**; refer to Figure-21.

Figure-20. Applications tab with the Apps related to Manufacturing

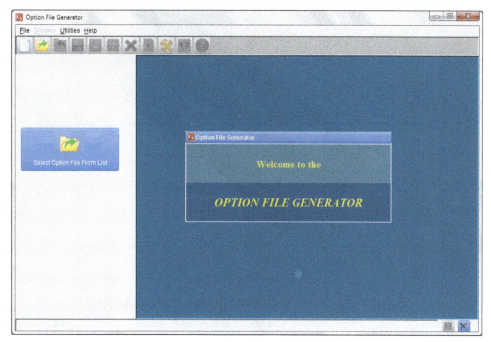

Figure-21. The Option File Generator App

The options and their use are not in the scope of this book.

TUTORIAL

In this tutorial, we will open the file created in Tutorial of Chapter 5. Now, we will create tool path and then the output file for this tutorial; refer to Figure-22 and Figure-23.

Figure-22. Model for tutorial

Figure-23. Backside of the model

Copying and Opening the tutorial file

- Copy the manufacturing assembly files created in Tutorial of Chapter 5.
- Create a new folder with the name **C06** and paste the files in this folder.
- Start Creo Parametric and open the c05 tut1.asm assembly file by using **Open** tool of the **Ribbon**.
- By default Manufacturing app of Creo will open. Now, click on the Applications tab in the Ribbon and select the Expert Machinist button from the Ribbon; the file will open in Expert Machinist App.

In this file, we have created various NC features along with their respective operations. Now, we will generate the tool paths and outputs of each operation consisting of various nc features.

Note that in this tutorial, tool path creation and output will be explained for only one feature and the rest work is to be done by you.

Creating Tool path for Profile feature

- Select PROFILE1[OP010] from the **Model Tree** and right-click on it. A shortcut menu will be displayed.
- Select the **Create Toolpath** option from the menu, the **Profile Milling** dialog box will be displayed as shown in Figure-24.
- In this dialog box, click on the down arrow next to **Cutting Tool**; a list of tools created earlier will be displayed.

Figure-24. Profile Milling dialog box

- Select the first tool from the list i.e. 1:END MILL ROUGH. You can preview the tool by selecting **Show Tool** button displayed adjacent to the down arrow.
- Now, we will specify the depth of cut in this operation. Click in the **Rough to** edit box in the dialog box and specify the value as 1.
- Now, we will specify the tool motion while cutting material. Select **Back and Forth** radio button from the dialog box so that the tool cuts material in both direction (forward and backward). Note that the tool type is an important factor for back and forth cutting.
- If the tool being used in machine is of very good quality then you can select the **Stay in cut** radio button from the dialog box. On selecting this radio button the tool will remain engaged with the material.
- If the parameters related to feed and spindle speed have not been defined for the machine then select the **Tool Path Properties** button from the dialog box; **Tool Path Properties** dialog box will be displayed as shown in Figure-25.

Figure-25. Tool Path Properties dialog box

- Specify the spindle speed for **Rough Speed** and **Finish Speed** as **2500** and **2800**, respectively.
- Click on the **Feed Rates** tab and specify the desired feed rate values in the corresponding edit boxes.
- Click on the **Cut Control** tab and specify the value of **Depth of cut** and **Stepover** as 2.
- Click on **OK** button from the dialog box and then OK from the **Profile Milling** dialog box, the tool path will be created and added in the Model Tree with the name PROFILE1 TP1[010].
- Click two times with a pause on its name and specify the name as Profile rough.

Now, we will create tool path to finish profile milling operation.

- Right click on **PROFILE1[OP010]** in the Model Tree and select the **Create Toolpath** option from the menu displayed, **Profile Milling** dialog box will be displayed as shown in Figure-26.
- In **Tool Path Name** edit box, specify the name of tool path as **Profile Finish**.
- Select **END MILL FINISH** from the **Cutting tool** list.
- Clear the **Rough** Check box and select the **Finish** check box. Now, select the **Finish Cuts** button, **Finish Cuts** dialog box will be displayed; refer to Figure-27.

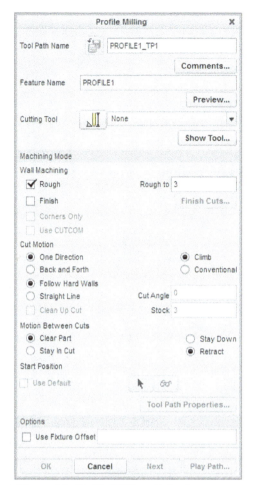

Figure-26. *Profile Milling dialog box*

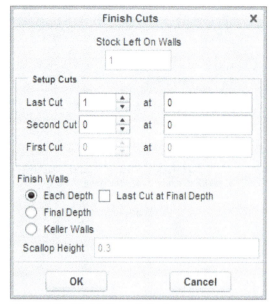

Figure-27. *Finish Cuts dialog box*

- Set the value as **2** in the **Last Cut** spinner and select the **Last Cut at Final Depth** and then select the **OK** button from the dialog box.
- Rest of the properties will remain unchanged. Now, select the **OK** button from the **Profile Milling** dialog box; the finishing tool path will be created.

Now, we will generate the output of the tool paths.

Generating tool path output

- Select the **PROFILE ROUGH** from the Model Tree and click on the down arrow adjacent to G00 button ⚙ displayed in the **Output** panel of the **Machining** tab in the **Ribbon**. A list of tools will be displayed.
- Click on the **Select Post** tool in this list, **Save a Copy** dialog box will be displayed.
- Save the file in the desired directory. On saving the file, **PP List** Menu Manager will be displayed with the list of Post Processor available in Creo.
- Select **UNCX01.P20** from the list and the output will be created in the directory in which you earlier saved the file using **Save a Copy** dialog box. Also, INFORMATION WINDOW will be displayed with the report of errors and warnings if any; refer to Figure-28.

Figure-28. INFORMATION WINDOW

- Similarly, you can create output for the PROFILE FINISH [OP010] toolpath.

In the same way, you can create tool path and output for the other operations. Note that manufacturing wise all the operations do not require roughing and finishing move. It mainly depends on the tolerance and surface finish in which the output is desired.

PRACTICE 1

In this problem, you will open the file created in Problem 1 of Chapter 5. In this file, you will generate all the tool paths and then you will create the outputs for UNCX01.P43 Post Processor. Figure-29 and Figure-30 show the model.

Figure-29. Isometric view of the model

Figure-30. Reverse side view of the model

PRACTICE 2

In this problem, you will open the file created in Problem 2 of Chapter 5. In this file, you will generate all the tool paths and then you will create the outputs for UNCX01.P43 Post Processor Figure-31 and Figure-32 show the model.

Figure-31. View for Problem 2

Figure-32. View for Problem2

Chapter 7

CMM Introduction

Topics Covered

The major topics covered in this chapter are:

- *Introduction to CMM and its purpose*
- *CMM Machine Setting*
- *Use of default probe library and creation of probe model*
- *Creation of measuring operation*
- *Various options for measurement*
- *Creating output file*

INTRODUCTION TO CMM AND ITS PURPOSE

CMM is the short form of Coordinate Measuring Machine. As the name suggests that this machine is used to measure coordinates. Now, question is which coordinates and why? One of the good reason of using CMM is accuracy required to manufacture parts in today's world. The curvatures and shapes created by NC machines are very complicated and are required to be accurate. But how to measure that accuracy? So, for measuring complicated shapes; CMM was introduced in production technology. A CMM works in the same way as does the NC machine. But, in CMM, in place of tool there is a probe that runs over the workpiece and gives coordinates of each desired point on the workpiece. These coordinates together generate point cloud which can be converted to surface. This generated surface is then compared to the standard size required and the dimensional and geometric analysis is performed. In a simple way, a CMM finds out the difference between the coordinate sheet specified by the user and the real model. The CMM app of Creo Parametric is used to generate a sheet of coordinates that will be fed to the machine to perform dimensional and geometric analysis on the real model. Since working with CMM app is similar to the Expert Machinist so the workflow for CMM will be the same. The process for generating Coordinates by this app is given the steps discussed next.

Note that in this book we will cover only the introduction to CMM so we will be discussing about the steps that are mandatory for generating Coordinates for CMM.

Adding Reference Model

This is the first step to setup coordinate sheet.

• To add a reference model, click on the down arrow displayed below Reference Model tool in the **Components** panel of the **Ribbon**. On clicking a list of tools is displayed; refer to Figure-1.

Figure-1. Add Reference Model drop-down

• Select the **Assemble Reference Model** option. The **Open** dialog box will be displayed, as shown in Figure-2.

Figure-2. Open dialog box

- Select a model from the dialog box and then click on the **Open** button from dialog box. The model will be displayed in Modeling area and you will be prompted to assemble the part with the assembly plane.
- Select the **Default** option from the **Constraints** drop-down; refer to Figure-3. The reference model will be assembled at the default planes.

Figure-3. Constraints drop-down with Default option highlighted

- Click on the **Done** button ✓ to accept the assembly, the reference model will be placed.

Setting Up a Fixture

After adding the reference model, you can add a fixture to hold the reference model. This step is optional and is perfomed when you want to check the interference of the probe with the fixture. The procedure to add a fixture is discussed next.

• Click on the **Fixture** button 🔧 in the **Components** panel in the **Ribbon**, the Fixture Setup contextual tab will become active.
• Here, you can set the models to be used as fixture for our current reference model. To add model for the Fixture, click on the **Components** tab in the **Ribbon**; the tab will be displayed in expanded form as shown in Figure-4.

Figure-4. Expanded Components tab

• In this tab, click on the **Add a Fixture Component** button, the **Open** dialog box will be displayed again.
• Using the options in this dialog box, you can add an assembly model or part file as fixture in the setup.

CMM Setting

Till this point, we have a reference model and hopefully a fixture in the modeling area. Now, we need to setup a CMM so that the coordinate data to be generated is according to our CMM. The procedure to set up CMM is discussed next.

• Click on the down arrow displayed below **CMM** in the **Inspection Machine** panel of the **Ribbon**. A list of tools will be displayed as shown in Figure-5.

Figure-5. CMM flyout

• Click on the **CMM** tool from the list. The **CMM Work Center** dialog box will be displayed; refer to Figure-6.

Figure-6. CMM Work Center dialog box

- The options in this dialog box are similar to the options discussed for **Milling Work Center** dialog box in Chapter 3.
- The main difference between the two dialog boxes is that; in **Milling Work Center** dialog box, a tool is required and in **CMM Work Center** dialog box, a probe is required in place of tool.

The options related to probe in this dialog box are discussed next.

Probe

A probe is a solid and sensitive tool which is used to map the surface of a component in the machine.

- To add a probe in the setup, select the **Probe** button in the **Probes** tab of the **CMM Work Center** dialog box; refer to Figure 7-6. On selecting this button, the **Probe Setup** dialog box will be displayed; refer to Figure-7.

Figure-7. Probe Setup dialog box

- Now, you need to add an existing probe in the setup by using the **Open** button. On selecting the **Open** button, you will be asked to select model or assembly of the probe.
- You can also use the default probes provided by Creo. To access the default probes, click on the **File** button in the menu bar of the **Probe Setup** dialog box; a menu will be displayed as shown in Figure-8.

Figure-8. File menu

- Select the **Default Probes** button, the **Open** dialog box will be displayed with the folders of defaults probes in it; refer to Figure-9.

Figure-9. The Open dialog box with the default probes in folders

- Open the desired folder in the dialog box and select the assembly file from it. In our case, the **indx star tip** folders is opened and the assembly file in it has been selected.
- Now, click on the **Open** button from the dialog box. The **Probe Setup** dialog box will change as shown in Figure-10.

Figure-10. Probe Setup dialog box after selecting default probe

- Specify the desired settings in this dialog box and click on **Apply** button.
- Now, click on the **OK** button from the dialog box. The probe will be added in the setup and the **CMM Work Center** dialog box will be displayed again.

Now, specify rest of the settings for the CMM Work Center and then click on **OK** button from the dialog box, the machine will be added in the **Model Tree**.

Creating Measuring Operations

As in Expert Machinist, we create an operation first to create tool path sequence, in the same way we create operation in CMM app before creating measurement sequences. The procedure is given next.

- Select **Operation** tool from the **Process** panel in the **Ribbon**, the **Operation** contextual tab will be displayed in the **Ribbon** as shown in Figure-11.

Figure-11. Operation contextual tab

- The CMM earlier created is already selected in **Work Center** drop-down in the **Ribbon**.
- Now, select a coordinate system from the modeling area. Note that the coordinate selected will be assumed as **0,0,0** position while measuring.
- Select the **Done** button ✔ from the **Ribbon** to create the operation. After creating the operation most of the tools in the **Ribbon** will become active.

Creating Measuring Steps

Measure steps in CMM app work in the same way as do the NC features in Expert Machinist app. The only difference is that in this case, you will be measuring the dimensions and geometry in place of cutting material. In this introductory chapter, only **Plane** measure step is explained. The other measure steps also work in the same way. Creation of **Plane** measure step is explained next.

- After setting up probe, CMM and operation; click on the **Plane** tool available in the **Measure** panel of the **Ribbon**. The Probe setup dialog box is displayed.
- On clicking **OK** button, The **Edit Parameter** dialog box is displayed; refer to Figure-12.

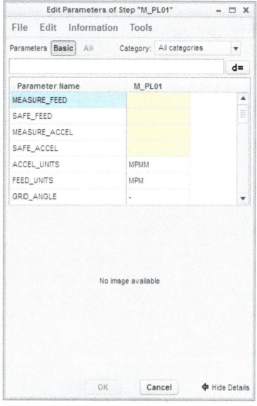

Figure-12. Edit Parameter dialog box

- Now, specify the parameters for measurement in the fields of this dialog box. The fields that are highlighted in yellow color are compulsory to specify and the other fields are optional.
- After specify the desired parameters, click on the **OK** button from the dialog box.
- Now, you will be asked to select the planar face which is to be used as reference for CMM measurements; refer to Figure-13.

Figure-13. Face to be selected with cursor on it

- On selecting the face **PTS and PATH** Menu Manager is displayed.
- Select **Automatic** option from the **ADD POINTS** Menu Manager displayed at the bottom in the **PTS and PATH** Menu Manager. The **Probe Points** dialog box is displayed as shown in Figure-14.

Figure-14. Probe Points dialog box

- Increase the number of points to measured by probe in both U and V directions by using their respective spinners in this dialog box.
- Click on the **OK** button from the dialog box. Preview of the probe points will be displayed in the modeling area; refer to Figure-15.

Figure-15. Preview of the probe points

- Select the **Done/Return** option from the **ADD POINTS** Menu Manager and **PTS and PATH** Menu Manager, the probe points will be added in the setup.
- Select the **OK** button from the **MEASUREMENT STEP** dialog box displayed in the left of the application window; refer to Figure-16.

Figure-16. MEASUREMENT STEP dialog box

- On selecting the **OK** button, the step is created and added in the Model Tree (and highlighted in blue color in the modeling area); refer to Figure-17.

Figure-17. Measurement step created

In the same way, you can create other measurement steps like surface, cylinder, cone, and so on.

Probe Path Output

After creating a measurement step, this is the time to generate output for the machine. The tool to generate output is available in the **Output** panel with the name as **Probe Path**.

- Select the **Probe Path** tool from the **Ribbon**. The **PROBE PATH** Menu Manager is displayed.
- Now, you can create output for a complete operation having various measurement steps or you can create output for a specific step by using the options in this Menu Manager. The procedure for both the cases will be the same.
- To generate output for individual measurement step, select **CMM Step** from the Menu Manager. On doing so, list of available measurement steps will be displayed; refer to Figure-18.
- Select the step from the list displayed. The Menu Manager will get changed as shown in Figure-19.

*Figure-18. PROBE PATH
Menu Manager*

Figure-19. Modified PROBE PATH Menu Manager

• Select **File** option from the Menu Manager to generate the output file, the **Save a Copy** dialog box will be displayed.
• Save the file at the desired location. After saving the file, select **Done** option from the Menu Manager.
• The Menu Manager will display two options: **Play Path** and **Show File**.
• You can check the simulation of probe path by using the **Play Path** option.
• Using the **Show File**, you can check the output file that you have saved.
• After selecting any of the two options (**Play Path** and **Show File**), select **CMM Step** and then the desired step from the step list; the output will be displayed.

Tip

This is important to note that the CMM steps are used to perform geometric and dimensional analysis. So, make sure that you have applied geometric tolerances in the model as well.

PRACTICE 1

The models that we have earlier used can be check for CMM by using the tools discussed in this chapter. So, use the model files provided in the resource kit to check for CMM output.

PRACTICE 2

Use the model with the name C07Prob2.prt as reference model and then create CMM output for the model, refer to Figure-20.

Figure-20. Model for Problem 2

FOR STUDENT NOTES

Chapter 8

Project

Topics Covered

The major topics covered in this chapter are:

- *Starting an Expert Machinist File*
- *Importing the reference model and creating the workpiece*
- *Creating Machine Setup*
- *Creating Operation Setup*
- *Creating Tool Setup*
- *Creating NC Features*
- *Creating Tool paths*

PROJECT

Create NC program for the model shown in Figure-1. The workpiece for machining is a rectangular envelope of the model.

Figure-1. Model for project

Starting an Expert Machinist File

- Start Creo Parametric by using the icon on desktop or by using the Start menu.
- Select the **New** button from the **Data** panel in the **Home** tab. The **New** dialog box will be displayed.
- Select the **Manufacturing** radio button from the **Type** area and **Expert Machinist** option from the **Sub-type** area.
- Specify the name as **Project** in the **Name** edit box of this dialog box and clear the **Use default template** check box. After specifying all the settings, the updated **New** dialog box will be displayed; refer to Figure-2.

Figure-2. New dialog box

- Click on the **OK** button from the dialog box. The **New File Options** dialog box will be displayed; refer to Figure-3.

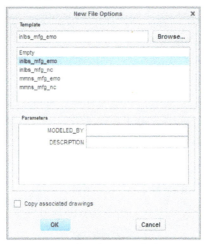

Figure-3. New File Options dialog box

- Select **mmns mfg nc** template from the list, enter your name in the **MODELED BY** box of this dialog box and then click on the **OK** button. On doing so the interface of Expert Machinist is displayed.
- Click on the **Quit NC-Wizard** button to exit the **NC-WIZARD** dialog box displayed in the right. This will increase the view area. You can skip this step if you want to see the tips related to current operation.

Importing the reference model and creating the workpiece

- Click on **Create Model** button from the **NC Model** panel of the **Machining** tab of the **Ribbon**; refer to Figure-4. The prompt will be displayed to specify the name of the nc model that will be used for machining; refer to Figure-5.

Figure-4. Create Model button

Figure-5. Prompt for specifying name of the nc model

- Click on the tick mark button to accept the default name displayed in as **[PROJECT NCMDL]** or you can specify your desired name then select the tick mark button. On select the button, the **Open** dialog box will be displayed along with the Menu Manager; refer to Figure-6.

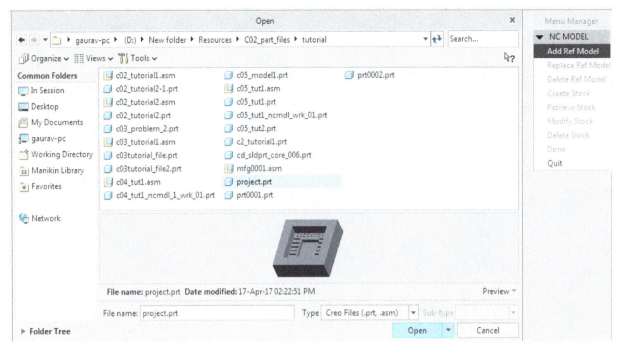

Figure-6. The Open dialog box with the NC MODEL menu manager

- Select the **project.prt** file from the downloaded resource kit and then click on **Open** button from the **Open** dialog box, the model will be displayed in the Expert Machinist interface; refer to Figure-7.

Figure-7. Expert machinist interface with the imported model and the active NC MODEL Menu Manager

- Click on **Create Stock** option from the Menu Manager, the imported model(also called reference model) will be packaged by a rectangular box; refer to Figure-8.

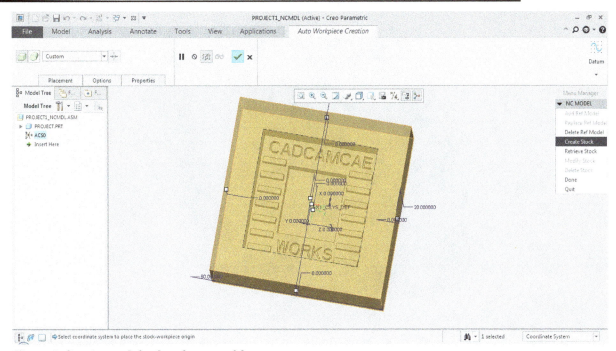

Figure-8. Creating stock for the reference model

- This rectangular box will act as workpiece layer. Click on the **Done** button ✔ from the **Auto Workpiece Creation** contextual tab displayed in the **Ribbon**.
- On selecting the **Done** button, the rectangular box will be locked as the workpiece. Select the **Done** button from the Menu Manager to accept the workpiece. On doing so, the **Component Placement** contextual tab will be displayed in the **Ribbon**.
- Select the **Default** option from the **Constraint** drop-down in this contextual tab, refer to Figure-9. The workpiece along with the reference model is placed at the origin of Expert Machinist coordinate system.

Figure-9. Option to be selected from Constraint drop-down

- Click on the **Done** button ✔ from the **Ribbon** to accept the placement.
- Click on **Material Removal Display** button from the **Display** panel of **Machining** tab in the **Ribbon** for a better visualization of workpiece. On selecting this button, the workpiece will be displayed as shown in Figure-10.

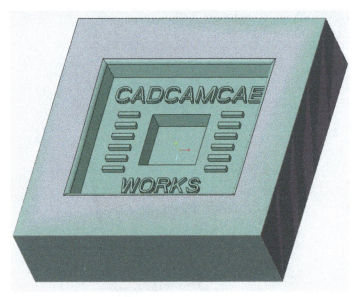

Figure-10. The workpiece displayed after selecting Material Removal Display button

Creating Machine Setup

- Select the **Machine Tool Manager** tool from the **Work Center** drop-down in the **Setup** panel of **Machining** tab in the **Ribbon, Milling Work Center** dialog box will be displayed as shown in Figure 8-10.
- Specify the name of machine as VMC015 in the **Name** edit box. (Now, the machine setup being created will be saved with the name VMC015)
- In the **CNC Control** edit box, specify the name of controller as Haas.
- Select the **5 Axis** option from the **Number of Axes** drop-down.
- In the **Output** tab of this dialog box, retain all the default settings as shown in Figure-11.

Figure-11. The Milling Work center dialog box

- Click on the **Parameters** tab to activate the options related to parameters of machine.
- In the **Maximum Speed (RPM)** edit box, specify the speed as **8000**.
- In the **Horsepower** edit box specify the value as **30**.
- Specify the rapid feed rate as **14000** in the **Rapid Feed Rate** edit box. Make sure that **MMPM** is selected in the **Rapid Traverse** drop-down in this tab. On specifying all parameters, the updated dialog box should be displayed as shown in Figure-12.
- Click on the **Travel** tab to specify the limits of machine.
- Specify the value of X travel as **700**, Y travel as **400**, and Z travel as **500**. Refer to Figure-13 for specifying the values.

Figure-12. Parameters tab of Milling Work Center dialog box

Figure-13. The Travel tab of Milling Work Center dialog box

- Click on the **OK** button to accept all the specified settings and create the machine. On doing so, the machine will be added in the **Model Tree** with the name **VMC015**.

Creating Operation Setup

- Click on the **Operation** tool from the **Setup** panel in the **Ribbon**. The options of **Operation** tab will be displayed in the **Ribbon**; refer to Figure-14.

Figure-14. Operation tab and its options

- Click on the **Work Center** drop-down and select the **VMC015** option. (the machine we created earlier)

- Click on the **Datum** drop-down (displayed in the right of the **Ribbon**) and select **Coordinate System** button ⚹, the **Coordinate System** dialog box will be displayed; refer to Figure-15.

Figure-15. Coordinate System dialog box

- Select the top face of the workpiece to place the coordinate system.
- Select the two boundary edges and specify the distance value as **0** for both; refer to Figure-16.

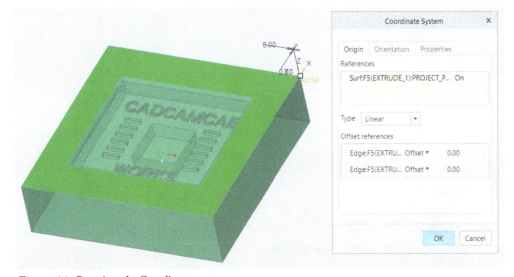

Figure-16. Creating the Coordinate system

- Click on the **OK** button from the **Coordinate System** dialog box. The coordinate system will be placed at the specified location.
- Click on the **Resume** button ▶ from the **Ribbon** to activate the **Operation** tab again.
- Select the coordinate system recently created and then click on the **Done** button from the **Ribbon** to create the operation.

Note that after creating the operation, the **Cutting Tool Manager** tool also becomes active in the **Ribbon**.

Creating Tool Setup

- Click on the **Cutting Tool Manager** tool from the **Ribbon**, the **Tools Setup** dialog box will be displayed as shown in Figure-17.

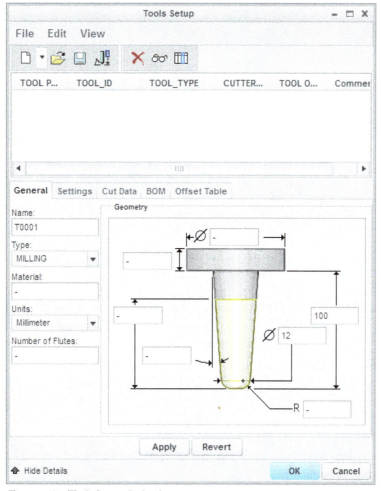

Figure-17. Tools Setup dialog box

- Click in the **Name** edit box of this dialog box and specify **END MILL ROUGH**.
- Click on the down arrow of **Type** drop-down and select the **END MILL** option.
- In the **Geometry** area of the dialog box, specify the values as shown in Figure-18.

Figure-18. Value to be specified

- Click on the **Cut Data** tab. The options in the tab will be displayed as shown in Figure-19.

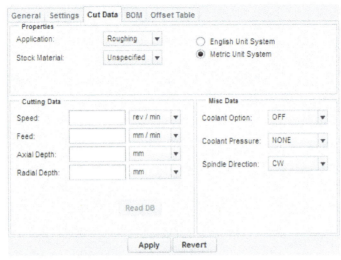

Figure-19. Options of the Cut Data tab

- Click in the **Speed** edit box and specify the value as **4500**.
- Click in the **Feed** edit box and specify the value as **250**.
- Click in the **Axial Depth** edit box and specify the value as **2**.
- Click in the **Radial Depth** edit box and specify the value as **2**.
- Click on the down arrow of **Coolant Option** drop-down and select the **FLOOD** option from the list.
- Click on the down arrow of **Coolant Pressure** drop-down and select the **MEDIUM** option from the list.
- Click on the down arrow of the **Spindle Direction** drop-down and select the **CCW** option from the list.
- Now, click on the **Apply** button to create the tool, the tool will be added in the tool list of **Tools Setup** dialog box.

For Finishing and embossing the text, we need to add a finishing tool also. Following steps can be used for creating the finishing tool.

- Click on the **General** tab if any other tab is selected.
- Click in the **Name** edit box and specify the name as **END MILL FINISH**.
- Select **END MILL** option from the **Type** drop-down if not selected.
- Specify the parameters of the tool as shown in Figure-20.

Figure-20. Parameters to be specified for tool

• Click on the **Settings** tab, the options in the **Settings** tab will be displayed as shown in Figure-21.

Figure-21. Options in the Settings tab

• Click in the **Tool Number** edit box and specify the value as **2**.
• Click on the **Cut Data** tab to display options related to cutting; refer to Figure-22.

Figure-22. Options in the Cut Data tab

• Click on the down arrow of **Application** drop-down and select the **Finishing** option from the list.
• Specify the rotational speed of tool as **5000** in the **Speed** edit box.
• Click on the down arrow of **Coolant Pressure** drop-down and select the **MEDIUM** option from the list.
• Specify the rest of the options as in Figure-22.

Now, click on the **Apply** button to add the tool in the list. The reference part also has fillet at the corners. To create fillet in the workpiece, we need a special type of tool. The procedure to add this tool is given next.

• Click on the **General** tab if not selected.
• Click in the **Name** edit box and specify **FILLET TOOL**.

- Click on the down arrow of **Type** drop-down and select the **CORNER ROUNDING** option from the list; the tool and its parameters will be displayed accordingly.
- Specify the parameters of the tool as shown in Figure-23.

Figure-23. Parameters to be specified for the tool

- Click on the **Settings** tab and specify the tool number as 3 in the **Tool Number** edit box.
- Click on the Cut Data tab, the options to specify cutting parameters will be displayed.
- Specify the cutting parameters as shown in Figure-24.

Figure-24. Specifying parameters for Fillet tool

- Click on the **Apply** button to add the tool in the list.

- Now, click the **OK** button from the **Tools Setup** dialog box to create the tool setup. On doing so, the options in the **NC Features & Machining** panel will become active.

Now, we are ready to create material cutting features (NC Features) to remove the material.

Creating NC Features

In this section, we will create the nc features to remove material from the workpiece. Selection of the nc feature is mainly based on the interpretation of user. There can be various ways to remove the same material. The selection of features will also be dependent on type of machinery being used and the tool setup. The NC Features used in the project are for ideal conditions when tools and machinery are available as per the requirement.

Round Feature

- Select the **Top Round** tool from the list displayed on clicking at the down arrow displayed next to **Top Chamfer** tool in **NC Features & Machining** panel in the **Ribbon**, the **TOP ROUND** dialog box will be displayed along with **SELECT SRFS** Menu Manager.
- Select the round faces of the workpiece to create the nc feature; refer to Figure-25 and then press the MMB (Middle Mouse Button) twice.

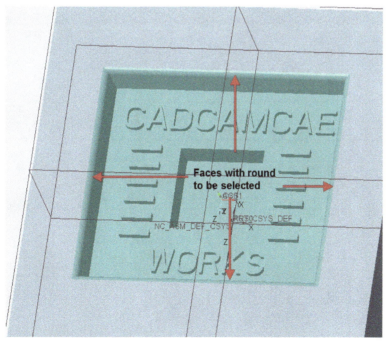

Figure-25. Faces to be selected for Top Round feature

- Now, the **OK** button of the **TOP ROUND** dialog box will become active. Click on the **OK** button from the dialog box, the round feature will be created.

Pocket Feature

- Click on the **Pocket** tool from the **Ribbon**, the **Pocket Feature** dialog box will be displayed along with the **SELECT SRFS** Menu Manager.
- Select the bottom face of the reference model; refer to Figure-26.

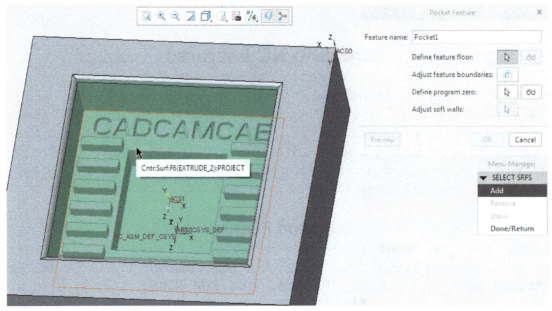

Figure-26. Face to be selected for Pocket feature

- Press the MMB twice. The **Define/Adjust Feature Top** dialog box will be displayed.
- Click on the **Cancel** button as we do not want to change location of feature top. Preview of the pocket feature will be displayed as shown in Figure-27.

Figure-27. Preview of pocket feature

At some places the feature will not be according to our requirement, so we need to modify the boundaries of the feature.

- Select **Adjust Feature Boundaries** button from the dialog box, the sketcher environment of Creo will open; refer to Figure-28.

Figure-28. Pocket feature boundaries in sketcher

Machining of the text requires high finish and cannot be achieved by our roughing tool. So, we need to exclude it from the current pocket feature.

- Select the boundaries around the text by using the window selection method; refer to Figure-29 and delete them.

Figure-29. Boundaries to be selected

- Change the display style to **Hidden Line** by using the In-Graphics Toolbar; refer to Figure-30. On doing so, the model will be displayed as shown in Figure-31.

Figure-30. In-Graphics Toolbar with the option to be selected

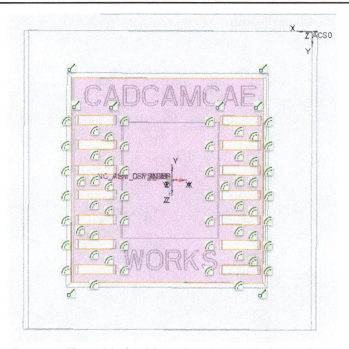

Figure-31. The model after deleting boundaries and selecting the Hidden line style

- Delete the outer boundary of the pocket feature and project the lower edges of fillet as boundary lines. Also, create rectangle around the text; refer to Figure-32.

Figure-32. Boundary sketch to be created

- Click on tick button from the **Ribbon** and then **OK** button from the **Pocket Feature** dialog box. The workpiece after switching to **Shading** display style will be displayed as shown in Figure-33.

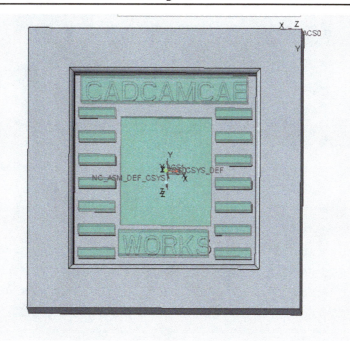

Figure-33. The workpiece after creating pocket feature

Now, we need to create one more pocket feature to remove material from the inner pocket.

- Click on the **Pocket** tool from **NC Features & Machining** panel in the **Ribbon**, the **Pocket Feature** dialog box will be displayed.
- Select the face as shown in Figure-34 and press the MMB twice.

Figure-34. Faces to be selected

- Select the **Adjust Feature Boundary** button from the dialog box, the sketcher environment will be displayed.
- Delete the older boundary and project the top boundary of the cavity; refer to Figure-35.

Figure-35. Projected top boundary of the cavity

- Click on tick button from the **Ribbon** and then click on the **OK** button from the **Pocket Feature** dialog box.

Slab Feature

- Select the **Slab** tool from the list displayed on clicking the down arrow next to **Flange** tool in the **Ribbon**. The **Slab Feature** dialog box will be displayed along with the **SELECT SRFS** Menu Manager.
- Select top face the rib feature; refer to Figure-36.

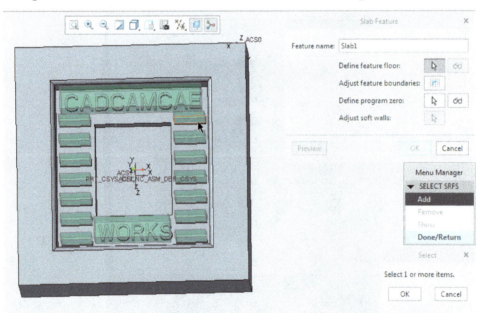

Figure-36. Face to be selected for slab feature

- Press the MMB twice and then click on the **Adjust Feature Boundaries** button from the **Slab Feature** dialog box. The sketcher environment will be displayed; refer to Figure-37.

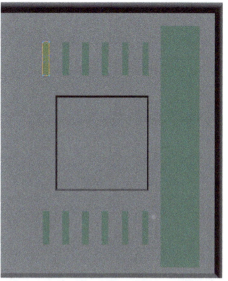

Figure-37. Sketcher environment for slab feature

- Select the **Project** tool from the **Ribbon** and project all the material boundaries by selecting them; refer to Figure-38.

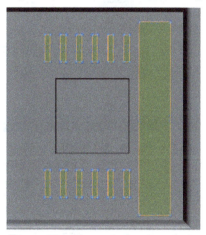

Figure-38. After projecting boundaries

- Select the **OK** button from the **Ribbon** and then from the **Slab Feature** dialog box, the model will be displayed as shown in Figure-39.

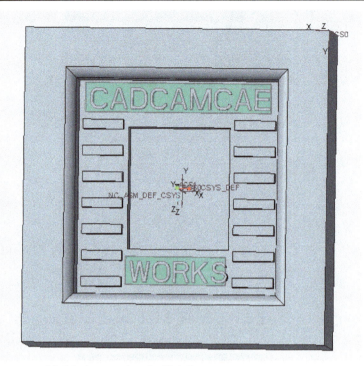

Figure-39. Model after creating the Slab feature

Now, we need to engrave the text on the workpiece. To do so, we need to create the step feature. The next steps explain the method of creating step feature.

Step Feature

- Select the **Step** tool from the **Ribbon**, the **Step Feature** dialog box will be displayed along with the **SELECT SRFS** Menu Manager.
- Select the face as shown in Figure-40. Press the MMB twice.

Figure-40. Face to be selected for Creating step feature

- Click on the **Adjust Feature Boundaries** button from the **Step Feature** dialog box, the sketcher environment will be displayed.
- Zoom in and project the inner boundaries of character 'A', 'O', 'R' and 'D'; refer to Figure-41.

Figure-41. Characters after projecting

- Click on the tick button from the **Ribbon** and then **OK** from the **Step Feature** dialog box.

The NC feature creation is complete till this point. Now, we need to generate tool path for the NC features.

Creating Toolpaths

In this section, we will create the toolpath to remove material from the workpiece by using respective nc features.

Toolpath for Round Feature

- Select **TOPROUND1 [OP010]** from the Model Tree and right click on it, a shortcut menu will be displayed.
- Select the **Create Toolpath** option from the list, the **Top Round/Chamfer Milling** dialog box will be displayed as shown in Figure-42.

Figure-42. Top Round/Chamfer Milling dialog box

- Click on the down arrow of **Cutting Tool** list and select the **3 : FILLET TOOL** from the list.
- Click in the **Stock** edit box and specify the value as **0**.
- Click on the **Tool Path Properties** button, the **Tool Path Properties** dialog box will be displayed; refer to Figure-43.
- Select the **From Tool** radio button for both **Rough Speed** and **Finish Speed** in the **Spindle Statements** area of the dialog box.
- Select the **Flood** radio button from the **Coolant** Options.
- Click on the **Feed Rates** tab, the dialog box will be displayed as shown in Figure-44.

Figure-43. Tool Path Properties dialog box

Figure-44. Tool Path Properties dialog box with the Feed Rates tab selected

- Click on the down arrow displayed below **Rough Cutting** and select the **From Tool** option from the list displayed.
- Select the **OK** button from the **Tool Path Properties** dialog box and then from the **Top Round/Chamfer Milling** dialog box, the tool path will be created and will be added below **TOPROUND1 [OP010]** NC feature in the Model Tree.
- You can play the tool path simulation by selecting the **Tool Path Player** option from the shortcut menu. This shortcut menu is displayed on right-clicking on the tool path in the Model Tree.

Toolpath for Pocket Feature

- Select **POCKET1 [OP010]** from the Model Tree and right click on it, a shortcut menu will be displayed.
- Select the **Create Toolpath** option from the list, the **Pocket Milling** dialog box will be displayed as shown in Figure-45.

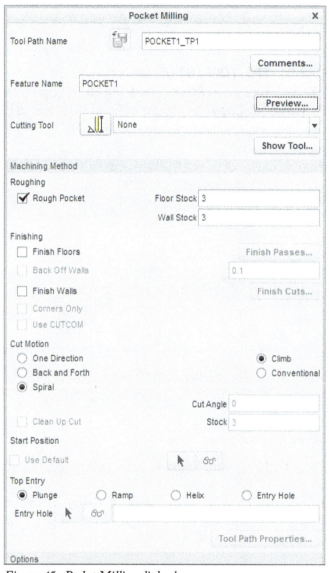

Figure-45. Pocket Milling dialog box

- Click on the down arrow of **Cutting Tool** list and select the **1 : END MILL ROUGH** from the list.
- Specify the value as **0** in both the **Floor Stock** and the **Wall Stock** edit boxes.
- Set the tool path properties similar to those specified round feature.
- Choose the **OK** button from the dialog box to create the tool path for the pocket milling feature.

Similarly, you can create tool paths of other features. Make sure that for the **Step1 [OP010]** nc feature, you select **2: END MILL FINISH** tool from the **Cutting Tool** list.

Creating Output Files

Till this point, we have simulated the tool path and we have checked it for errors. Now, we need to generate the output files by using the tool paths. In this section, we will generate the output files by using respective tool paths. To create an output file specific to a machine, we need to have a post processor specific to that machine. In this project, we will be using UNCX01.P12 as the post processor for our machine.

• Select the **TOPROUND1 TP1 [OP010]** from the Model Tree and then select the **Select Post** button from **Post Tool Path** drop-down of **Output** panel from **Ribbon**, the **Save a Copy** dialog box will be displayed; refer to Figure-46.

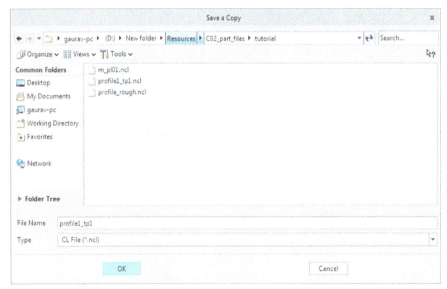

Figure-46. Save a Copy dialog box

• Save the Cl file at the desired location by using the options in this dialog box.
• On selecting **OK** button from the **Save a Copy** dialog box, the **PP LIST** Menu Manager (Post Processor List Menu Manager) will be displayed; refer to Figure-47.

Figure-47. PP LIST
Menu Manager

- Select the **UNCX01.P12** option from the Menu Manager, the **INFORMATION WINDOW** will be displayed and the output file will be saved at the location earlier specified in **Save a Copy** dialog box.
- Select the **Close** button from the **INFORMATION WINDOW**.
- To check the output, open the location specified earlier and then open the output file with .tap extension by using any word processor; refer to Figure-48.

Figure-48. Output file opened in Microsoft WordPad

- In the same way, you can generate and check the output files for the other tool paths.

Part II- NC Assembly

Chapter 9

Starting with NC Assembly

Topics Covered

The major topics covered in this chapter are:

- *Setting Reference Model*
- *Inherit Reference Model*
- *Automatic Workpiece*
- *Assemble Workpiece*
- *Workpiece Wizard*

INTRODUCTION TO NC ASSEMBLY

Creo Parametric 4.0 **NC Assembly** application creates the data necessary to drive an NC machine tool to machine a part for desired purpose. This application provide the tools to let the manufacturing engineer follow a logical sequence of steps to progress from a design model to ASCII CL data files that can be post-processed into NC machine data.

STARTING NC ASSEMBLY

There are two ways to start **NC Assembly** application which are discussed next.

Using New button

- Start the **Creo Parametric 4.0** if not started yet.
- Click on the **New** button from **File** menu of Creo Parametric window. The **New** dialog box will be displayed; refer to Figure-1.

Figure-1. The New dialog box

- Select the **Manufacturing** button from **Type** area and select **NC Assembly** from **Sub-Type** area.
- Click in the **Name** edit box and enter the desired name of file or part.
- Click in the **Common name** edit box and enter the description about the part. This option is optional.
- Clear the **Use default template** check box, if you want to set the template according to your need.
- Now, click **OK** button from **New** dialog box. The **New File Options** dialog box will be displayed; refer to Figure-2, where you need to select the specific template for your part.

Figure-2. New File Options dialog box

- Select the desired template from **Template** section.
- If you want to add more template to **Template** section then, click on the **Browse** button. The **Choose Template** dialog box will be displayed; refer to Figure-3.

Figure-3. The Choose Template dialog box

- Select the desired template and click on **Open** button. The new template will be added in **Template** section of **New File Option** dialog box.
- Click on the **MODELED BY** edit box of **Parameters** section and enter the detail of creator of this part. You can also add description in the **DESCRIPTION** edit box. These two options are optional.
- Select the **Copy associated drawings** check box, if you want to copy all the associated drawing files for the selected template. Otherwise clear the check box.

- After specifying the parameters, click on **OK** button from **New File Options** dialog box. The **NC Assembly** application window will be displayed; refer to Figure-4.

Figure-4. The NC Assembly application window

Using NC button

While working on part file, there is also an option to open the **NC Assembly** application which is discussed next.

- Click on the **NC** button of **Ribbon** from **Application** tab; refer to Figure-5. The NC Assembly window will be displayed; refer to Figure-6.

Figure-5. The NC button

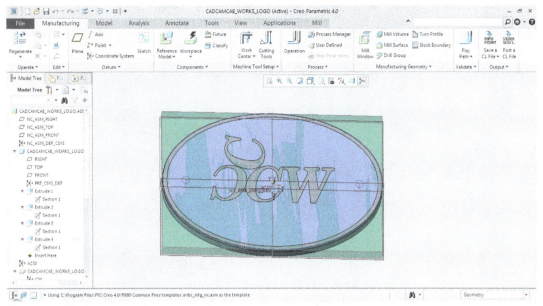

Figure-6. NC Assembly application window

- Now, you are ready to work on your part file in **NC Assembly** application.

SETTING MODEL AND WORKPIECE

In this section, we will discuss about the procedure of setting up a reference model and workpiece which are discussed next.

Setting up a reference model

- Click on the **Assemble Reference Model** button from **Reference Model** drop-down of **Component** panel; refer to Figure-7. The **Open** dialog box will be displayed; refer to Figure-8.

Figure-7. The Assemble Reference Model button

Figure-8. The Open dialog box for reference model

• Select the desired part file and click on **OK** button. The **Component Placement** tab will be displayed; refer to Figure-9. Using the options in this tab, you need to constrained the model.

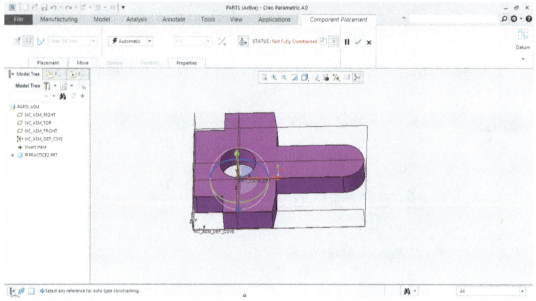

Figure-9. The Component Placement tab

• Click on **Default** button from **Constraint** drop-down of **Component Placement** tab to set the location of model to default position; refer to Figure-10. Please refer to Creo Parametric 4.0 Black Book for other constraining options.

Figure-10. Aassembling component to Default location

- On setting default location of part, the status will be changed from **Not Fully constrained** to **Fully constrained.**
- Click the **OK** button from the component tab to create the fully constrained model.

The other options of **Component Placement** tab are discussed next.

Component Placement tab

The **Component Placement** tab consists of commands, tabs, and shortcut menus. These commands, tabs, and shortcut menus are discussed next.

Constraint drop-down

This drop-down contains the constraints applicable for the selected model or set. By default, **User Defined** option is selected in the **Constraint Set** drop-down and **Automatic** is selected in the Constraint drop-down. The various constraint of **Constraint** drop-down are discussed next.

Automatic- Click on the **Automatic** button, if you want to constraint the model automatically based on your geometry selection.
Distance- Click on the **Distance** button, if you want to offset the component reference from the assembly reference.
Angle Offset- Click on the **Angle Offset** button, if you want to position the component at an angle to the assembly reference.
Parallel- Click on the **Parallel** button, if you want to orients the component reference parallel to the assembly reference.
Coincident- Click on the **Coincident** button, if you want to position the component reference coincident with the assembly reference.
Normal- Click on the **Normal** button, if you want to positions the component reference normal to the assembly reference.
Coplanar- Click on the **Coplanar** button, if you want to position the component reference coplanar to the assembly reference.
Centered- Click on the **Centered** button, if you want to center the component reference between the selected assembly references.
Tangent- Click on the **Tangent** button, if you want to position two references tangent to each other.
Fix- Click on the **Fix** button, if you want to fix the current location of a component that was moved or packaged.

Default- Click on the **Default** button, if you want to align the component coordinate system with the default assembly coordinate system.

⊿ If you want to flip the direction of offset then click on **Change Orientation of Constraint** button.

⊕ The **3D Dragger** button is used to hide the 3D Dragger from screen.

STATUS : Not Fully Constrained   The **STATUS** option is used to check the status of part or model. There are four types of status i.e. No Constraints, Not Fully Constrained, Fully Constrained and Constraints Invalid.
* No Constraints is displayed in **STATUS** section, when no constraints are applied to the model.
* Not Fully Constrained is displayed in **STATUS** section, when the model is partially constrained.
* Fully Constrained is displayed in **STATUS** section, when the model is fully constrained.
* Constrains Invalid is displayed in **STATUS** section, when the constraints of the model are invalid.

▣ The **Show component in separate window** button is used to display the component in its own window as you define constraints.

▣ The **Show component in assembly** window button is used to display the component in the graphics window and updates component placement as you define constraints.

Placement tab

The **Placement** tab is used to enable and display component placement and connection definition. There are various options in this tab which are discussed next.

Navigation and Collection area— This section displays **sets** and **constraints**; refer to Figure-11. Translation references and motion axes are displayed for predefined constraint sets. The first constraint in a set activates automatically. A new constraint activates automatically after a valid pair of references is selected, until the component is fully constrained.

Figure-11. Placement tab

* Select the **Constraint Enabled** check box to enable the Automatic option of **Set1** section. This **Automatic** option allow to set the reference.

- Select the desired constraint from **Constraint Type** drop-down.
- To set the reference you need to click on **Select Component Item** selection box and select the reference for component side.
- Click on **Select Assembly Item** selection box and select the assembly side reference for the model.
- On selecting the **Default** option of **Constraint** drop-down from **Component Placement** tab, the updated **Placement** tab will be displayed; refer to Figure-12.

Figure-12. Updated Placement tab

Move tab

The **Move** tab is used to move the component being assembled for easier access; refer to Figure-13. When the **Move** tab is active, all other component placement operations are paused.

Figure-13. The Move tab

- Select **Orient Mode** option from **Motion Type** drop down to reorient the view.
- Select **Translate** option from **Motion Type** drop-down to move the component.
- Select **Rotate** option from **Motion Type** drop-down to rotate the component.
- Select **Adjust** option from **Motion Type** drop-down to adjust the component position.
- Select the **Relative in view plane** radio button to move the component relative to the viewing plane.
- Select the **Motion Reference** radio button to move the component relative to a component or reference. On selecting this radio button reference selection is enabled where you need to select the reference from the screen.

- Click on the **Smooth** button of **Translation** drop-down to move/rotate the model smoothly..
- Click on the model and place the model wherever you want . While moving the model, the coordinate will be shown in **Relative** option.

After specifying the parameters for constrained model, click on the **OK** button from the **Component Placement** tab. The NC Assembly application window will be displayed with more enabled tools; refer to Figure-14.

Figure-14. Updated NC Assembly application window

Properties tab

The **Properties** tab display the information regarding the working model; refer to Figure-15.

Figure-15. Properties tab

- The **Name** box display the name of model.
- Click on the **Displays information for this feature** button to know more information about the current model.

Inherit Reference Model

The **Inherit Reference model** button is used to add some inherit feature. The procedure to use this button is discussed next.

- Click on the **Inherit Reference model** button of **Reference Model** drop-down from **Component** panel; refer to Figure-16. The **Open** dialog box will be displayed; refer to Figure-17.

Figure-16. The Inherit Reference Model button

Figure-17. The Open dialog box for reference model

- Select the file and click on the **Open** button. The **Component Placement** tab will be displayed, which was discussed earlier.
- Select **Default** option from **Constrained** drop-down and click on the **OK** button. The **Create Reference Model** dialog box will be displayed; refer to Figure-18.

Figure-18. The Create Reference Model dialog box

- Click in the **Name** edit box of **Reference Model** section and change the name if desired.
- Click in the **Common name** edit of **Reference Model** section and change the name as desired.

After specifying the parameters, click on the **OK** button to complete the process. The model will be added as manufacturing reference model.

Merge Reference Model

The **Merge Reference Model** tool is used to create a reference model with geometry merged from the selected model. The procedure to use this tool is discussed next.

- Click on the **Merge Reference Model** button of **Reference Model** drop-down from **Component Panel** in **Manufacturing** tab; refer to Figure-19. The **Open** dialog box will be displayed.

Figure-19. The Merge Reference Model button

- Select the part file from **Open** dialog box and click on the **Open** button. The **Component Placement** tab will be displayed which was discussed earlier.
- Select **Default** button from **Constrain** drop-down to fully constraind the model.
- Click on the **DONE** button from the **Component Placement** tab. The **Create Reference Model** dialog box will be displayed; refer to Figure-20.

Figure-20. The Create Refrence Model dialog box for merge by refeference

The options of this dialog box was discussed earlier.

After specifying the parameters, click **OK** button to complete the process.

CREATING WORKPIECE

The workpiece represents the raw material or stock that is going to be machined by the manufacturing operations. The use of workpiece is optional in Creo **NC Assembly** application. **NC Assembly** application allows you to create or assemble the workpiece. You can manually create the workpiece (manual workpiece) or allow **NC Assemby** to automatically create it using standard shapes and your choice of dimensions (automatic workpiece). Alternatively, you can assemble the workpiece using an existing design model. In this case, the workpiece can have features inherited or merged from the design model.

Automatic Workpiece

The **Automatic Workpiece** button is used to create workpiece automatically. The procedure to use this tool is discussed next.

- Click on the **Automatic Workpiece** button of **Workpiece** drop-down from **Components panel** in **Manufacturing** tab; refer to Figure-21. The **Auto Workpiece Creation** tab will be displayed; refer to Figure-22.

Figure-21. The Automatic Wworkpiece button

Figure-22. Auto Workpiece Creation tab

Auto Workpiece Creation tab

- Click on the **Create a rectangular workpiece** button, if you want to create a rectangular or billet shape workpiece.

- Click on the **Create a round workpiece** button, if you want to create a bar or round shape workpiece.

Sub-Shape drop-down

- Select the **Custom button** from **Sub-Shape** drop-down to define the overall dimensions and offsets of the workpiece; refer to Figure-23.
- Select the **Envelope** button from the **Sub-Shape** drop-down to define the minimum offsets and angular offsets. You can use **Envelope** button to reset the current offsets to the minimum values.

Figure-23. Sub-Shape drop-down

Placement tab

The **Placement** tab is used to specify the coordinate system and the reference model using their respective collectors.

- The **NC Assembly** application automatically specifies the coordinate system of the reference model in the **Coordinate System** edit box.
- If there is no reference model, or multiple reference models, or if you have removed the coordinate system, you need to specify it from the model.
- If you want to remove the coordinate system from selection then, right-click on the **Coordinate System** selection box and click on **Remove** button; refer to Figure-24.

Figure-24. The Placement tab

- The **NC Assembly** automatically specifies the reference model in the **Reference Model** edit box, if the manufacturing assembly has only one reference model.
- If the manufacturing assembly has multiple reference models, you need to specify the reference model for the workpiece.
- The procedure of **Remove** and **Information** button was discussed earlier but here you need to right-click on the **Reference Model** edit box.

Options tab

The **Options** tab is used to specify values for the units, overall dimensions, and stock offsets. The options in **Options** tab are different for rectangular and round workpiece.

On selecting the **Rectangular Workpiece** button, the **Options** tab is displayed as shown in; refer to Figure-25. The options for rectangular workpiece are discussed next.

Figure-25. The Options tab of rectangular workpiece

- Select **Millimeters** button from **Units** drop-down if you want to set the unit in millimeters.
- Select the **Inches** button from **Units** drop-down if you want to set the unit in inches.
- Click on the **X Total** edit box of **Overall Dimension** section and specify the overall dimensions for the workpiece in the x-direction. This value applies only for a billet workpiece.
- Click on the **Y Total** edit box of **Overall Dimension** section and specify the overall dimensions for the workpiece in the y-direction. This value applies only for a billet workpiece.
- Click on the **Z Total** edit box of **Overall Dimension** section and specify the overall dimensions for the workpiece in the z-direction. This value applies only for a billet workpiece.
- Click on the **Current Offset** button of **Linear Offset** section if you want to specify the stock offsets of the workpiece around the reference model and along the positive and negative x-axis y-axis, and z-axis.
- Click on the **Minimum Offset** button of **Linear Offset** section if you want to specify the minimum values of stock offsets of the workpiece around the reference model along the positive and negative x-axis, y-axis, and z-axis.

- Click on **About X** button of **Rotation offsets** section if you want to specify the angular stock offset of the workpiece around the x-axis of the reference model.
- Click on **About Y** button of **Rotation offsets** section if you want to specify the angular stock offset of the workpiece around the y-axis of the reference model.
- Click on **About Z** button of **Rotation Offsets** section if you want to specify the angular stock offset of the workpiece around the y-axis of the reference model.
- Click in the **Diameter Total** edit box and specify the diameter of the workpiece.
- Click in the **Length Total** edit box and specify the length of the workpiece.

Other options of the **Options** tab were discussed earlier.

Properties tab

The **Properties** tab is used to specify some information regarding the part file.

- Click in the **Workpiece Name** edit box and specify the desired name.
- Click in the **Workpiece Common Name** edit box and specify the desired name.

After specifying the parameters, click on the **Done** ✔ button from **Auto Workpiece Creation** tab. The workpiece will displayed; refer to Figure-26.

Figure-26. The NC Assembly application window after creating workpiece

Assemble Workpiece

The **Assemble Workpiece** button is used to assemble or add a workpiece with the existing workpiece. The procedure to use this button is discussed next.

- Click on the **Assemble Workpiece** tool of **Workpiece** drop-down from **Components** panel; refer to Figure-27. The **Open** dialog box will displayed.

Figure-27. Assemble Workpiece

- Select the part file which you want to add or assemble and the click on the **OK** button. The **Component Placement** tab will be displayed; refer to Figure-28.

Figure-28. Assembling the part file

- Now you need to constrain the newly added part with the existing part. Constrain the model as discussed earlier. (Refer to Creo Parametric 4.0 Black Book for more details).
- When the status is **Fully Constrained**, then click on the **Done** button. The **NC Assembly** application will be displayed; refer to Figure-29.

Figure-29. NC Assembly application window after adding part

Inherit Workpiece

The **Inherit Workpiece** tool is used for inherit features. The procedure to use this tool is discussed next.

- Click on the **Inherit Workpiece** tool of **Workpiece** drop-down from **Components** panel; refer to Figure-30. The **Open** dialog box will be displayed.

Figure-30. The Inherit workpiece button

- After selecting the part file for inherit feature, you need to click on the **Open** button. The **Component Placement** tab will be displayed which was discussed earlier.
- Click on the **Done** button from **Component Placement** tab. The **Create Stock-workpiece** dialog box will displayed; refer to Figure-31.

Figure-31. The Create Stock-workpiece dialog box

The options of **Create Stock-workpiece** dialog box were discussed earlier.

Merge Workpiece

The **Merge Workpiece** tool is used to merge features. The procedure to use this tool is discussed next.

- Click on the **Merge Workpiece** tool of **Workpiece** drop-down from **Component** panel; refer to Figure-32. The **Open** dialog box will be displayed.

Figure-32. Merge Workpiece tool

- After selecting the part file for merge feature, you need to click on the **Open** button. The **Component Placement** tab will be displayed which was discussed earlier.
- In this window, you need to fully constrain the part file with newly added part; refer to Figure-33.

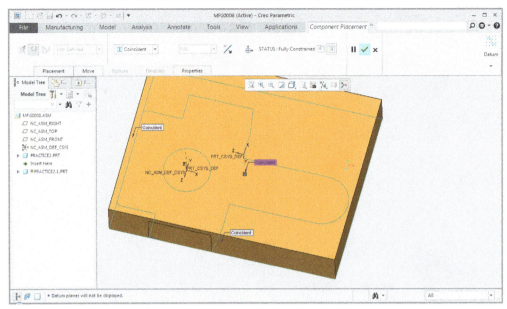

Figure-33. Fully Constrained part

- After specifying the parameters, click on the **OK** button. The **Create Stock-workpiece** dialog box will be displayed; refer to Figure-34.

Figure-34. The Create Stock-workpiece dialog box for merge feature

- The options of the **Create Stock-workpiece** dialog box were discussed earlier.
- After specifying the parameters, click on **OK** button. The **NC Assembly** application window will be displayed; refer to Figure-35.

Figure-35. The NC Assembly window after merge feature

Create Workpiece

The **Create Workpiece** tool is used to create a workpiece manually. The procedure to use this tool is discussed next.

- Click on the **Create Workpiece** tool of **Workpiece** drop-down from **Component** panel; refer to Figure-36. The **Enter Part name** edit box will be displayed on the application window.

Figure-36. The Create Workpiece tool

- Enter the name in the edit box or click on the **Done** button to use the default name of part. The **Menu Manager** will be displayed; refer to Figure-37.

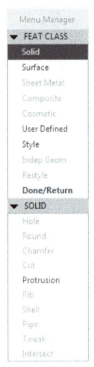

Figure-37. The Menu Manager

- You need to select the face of part; refer to Figure-38.

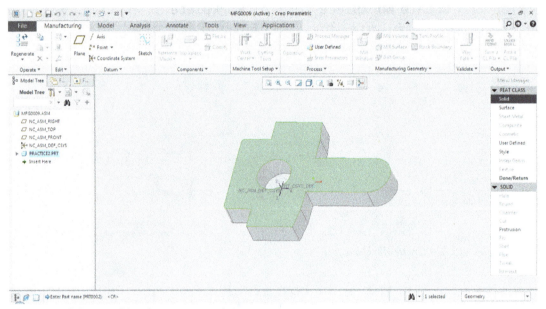

Figure-38. Selection of face for creating workpiece

- After selection of plane you need to click on **Protrusion** button of **Solid** section from **Menu Manager** dialog box. The Updated **Menu Manager** will be displayed; refer to Figure-39.

*Figure-39. Up-
dated Menu
Manager*

- Here, you need to select the type of workpiece you are going to make. Select **Extrude** and **Solid** button.
- After specifying the parameters, click on **Done** button from **Menu Manager** dialog box. The **References** dialog box will be displayed; refer to Figure-40.

Figure-40. Refrences dialog box

- In **References** dialog box, you need to add references to create the workpiece. In references you can add coordinate system or edges.
- After selecting the references, click on **Solve** button to check the references status. If the status is Fully Placed, then click on **Close** button.
- Now, create the sketch of workpiece as desired, then click on **OK** button from **Ribbon**. The **Extrude** tab will be displayed; refer to Figure-41.

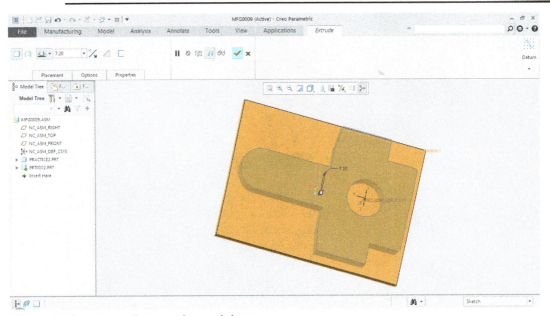

Figure-41. Creating workpiece with extrude button

- Adjust the dimension of extrude for workpiece as desired and click on **Done** button. The **NC Assembly** application window will be displayed along with part and recently created workpiece; refer to Figure-42.

Figure-42. The NC Assembly application window after Create Workpiece

Classify

The **Classify** tool is used to classify the model as reference, fixture, or workpiece. The procedure to use this tool is discussed next.

- Click on the **Classify** tool of **Components** panel from **Manufacturing** tab; refer to Figure-43. The **Select** dialog box will be displayed; refer to Figure-44.

Figure-43. Classify tool

- Select the model to create reference and press middle mouse button. The **MFG RECLASS Menu Manager** will be displayed; refer to Figure-45.

Figure-44. Select dialog box

Figure-45. MFG RECLASS Menu Manager

- Select **Ref Model** button from **Menu Manager,** if you want to create the selected model as reference model.
- Select **Workpiece** button from **Menu Manager,** if you want to create the selected model as workpiece.
- After selection of specific button, click on **Done** button. The model will be created as desired.

Component Operations

The **Component Operations** tool is used to apply different operations to the component. The procedure to use this tool is discussed next.

- Click on the **Component Operations** tool of **Component** panel drop-down from **Manufacturing** tab; refer to Figure-46. The **COMPONENT Menu Manager** will be displayed; refer to Figure-47.

Figure-46. Component Operations tool

Figure-47. COMPONENT Menu Manager

Copy button

The **Copy** button is used to copy component with respect to coordinate system. The procedure to use this tool is discussed next.

- Click on the **Copy** button from **COMPONENT Menu Manager.** The **GET COORD S** box will be displayed; refer to Figure-48.

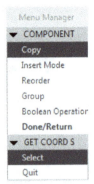

Figure-48. Updated COMPONENT Menu Manager

- Select the **Coordinate system** from the application window. The **Select** dialog box will be displayed; refer to Figure-49.

Figure-49. Selection of coordinate system for copy command

- Select the part to copy and press middle mouse button. The updated **Menu Manager** will be displayed; refer to Figure-50.

Figure-50. The Updated Menu Manager

- Select the transfer axis from **TRANS DIR** section. The **Input distance** box will be displayed on the canvas screen; refer to Figure-50.

- After inserting the distance, click on the **Done** button. The Updated **Menu Manager** will be displayed; refer to Figure-51.

Figure-51. Menu Manager

- Click on the **Done Move** button. The **Enter number of Instances** box will be displayed; refer to Figure-52.

Figure-52. Enter number of instances

- After inserting the value, click on the **Done** button. The multiple copy of selected model will be created; refer to Figure-53.

Figure-53. Creating multiple copy

Insert Mode

The **Insert Mode** tool is used to roll back assembly to insert one component before another. The procedure to use this tool is discussed next.

- Click on the **Insert Mode** button from **COMPONENT Menu Manager**. The **INSERT MODE Menu Manager** will be displayed.

Figure-54. IN-SERT MODE Menu Manager

- Click on **Activate** button. The **SELECT FEAT** box will be displayed.
- Now, click on any action or command from **Model Tree** to activate the insert mode; refer to Figure-55.

Figure-55. Model Tree for Insert Mode

- The commands or action below **Insert Here** button are hidden until **Insert Here** button is placed below all the commands.
- You can also do this process by drag and drop commands with the help of left key of mouse.

Reorder

The **Reorder** tool is used to reorder the sequence of component generation. The procedure to use this tool is discussed next.

- Click on the **Reorder** button from **COMPONENT** Menu Manager. The updated **Menu Manager** will be displayed along with **Select** dialog box; refer to Figure-56.

Figure-56. The Menu Manager for Reorder

- Select the part or action which you want to reorder. After selection click on **Done** button. The updated **Reorder Menu Manager** will be displayed; refer to Figure-57.

Figure-57. The updaterd Recorder Menu Manager

- Select **Before** or **After** option from **Reorder** section and click on the action or command for reordering from **Model Tree**.
- Click on the **Done** button. The action will be reordered.

Group

The **Group** tool is used to group components and assemble the features. The procedure to use this tool is discussed next.

- Click on the **Group** button of **COMPONENT** Menu Manager. The **Open** dialog box will be displayed along with updated **Menu Manager**; refer to Figure-58.
- Click on **Local Group** button from **Create Group** section. The **Enter Group Name** edit box will be displayed.
- Enter the desired name in **Enter Group Name** edit box and click on **Done** button. The Updated **Menu Manager** will be displayed along with **Select** dialog box; refer to Figure-59.

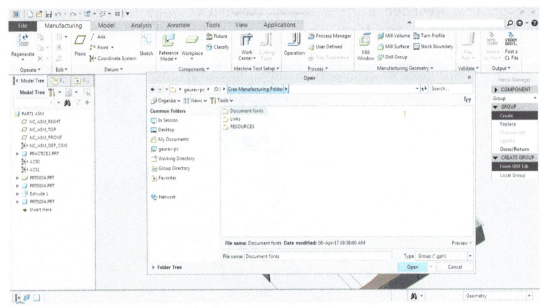

Figure-58. Updated Menu Manager for Group button

- Select the command or action while holding **CTRL** key from **Model Tree** and click on **Done** button from **Select Feat** section.
- The **Group** will be created in the **Model Tree**.

Boolean Operator

The **Boolean Operator** tool is used for boolean operation. The procedure to use this tool is discussed next.

- Click on the **Boolean Operator** button of **COMPONENT Menu Manager**. The **Boolean Operations** dialog box will be displayed along with updated **COMPONENT Menu Manager**; refer to Figure-60.

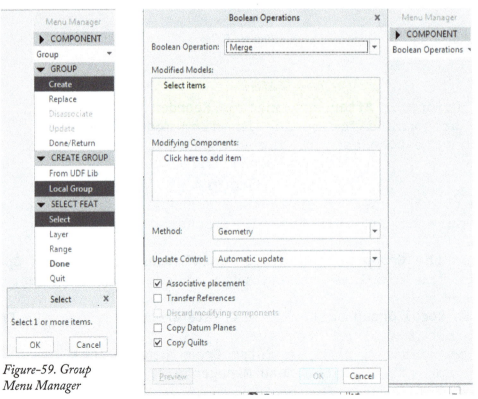

Figure-59. Group Menu Manager

Figure-60. The Boolean Operations dialog box

Boolean Operations drop-down

- Select **Merge** option from **Boolean Operations** drop-down of **Boolean Operations** dialog box, if you want to merge the selected component from selected model.
- Select **Cut** option from **Boolean Operations** drop-down of **Boolean Operations** dialog box, if you want to cut the selected component from selected model.
- Select the **Intersect** option from **Boolean Operations** drop-down of **Boolean Operations** dialog box, if you want to intersect the selected component from selected model.

Update Control drop-down

- Select **Automatic Updates** option from **Update Control** drop-down, if you want to update the feature automatically.
- Select the **Manual Updates** option from **Update Control** drop-down, if you want to update the feature upon request.
- Select **No Dependency** option from **Update Control** drop-down, if you do not want to create any dependency between a feature and the modifying component.
- Check the **Associative Placement** check box, if you want to update the resulting cut according to the placement of modifying component.
- Check the **Copy Datum Plane** check box, if you want to copy Datums to the modified model as a part of cut.
- Check the **Copy Quilts** check box, if you want to copy quilts to the modified model as a part of Cut.
- After specifying the parameters, click on the **Preview** button. If preview of your model is as per your need then click on the **OK** button from the **Boolean Operations** drop-down.
- The boolean operation will be created.

Manufacturing Assembly

The **Manufacturing Assembly** tool is used to assemble and classify the model. There are two tools in **Manufacturing Assembly** cascading menu which are discussed next; refer to Figure-61.

Figure-61. Manufacturing Assembly tool

Assembly and Classify

- Click on the **Assemble and Classify** tool of **Manufacturing Assembly** cascading menu from **Components** drop-down. The **Open** dialog box will be displayed.

- Select the assembly part file from **Open** dialog box and click on **Open** button. The **Component Placement** tab will be displayed.
- Set the constrain of recently added assembly to **Default** from **Constrain** drop-down. You can also constraint the model with assembly by yourself.
- Click on **Done** button from **Component Placement** tab. The **NC Assembly** application window will be displayed along with **MFG CLASS Menu Manager**; refer to Figure-62.

Figure-62. The NC Assembly MFG CLASS Menu Manager

- Select the **Ref Model** or **Workpiece** from **MFG CLASS Menu Manager** and select the model or part from reference as required.
- Note that you can even set some components of assembly as reference and other components of assembly as workpiece by using these options.
- Click on **Done** button from **Menu Manager** to complete the process.

Assemble

The **Assemble** tool is used to assemble a manufacturing assembly. The procedure to use this tool is discussed next.

- Click on the **Assemble** tool of **Manufacturing Assembly** cascading menu from **Components** drop-down. The **Open** dialog box will be displayed.
- Select the assembly part file from **Open** dialog box and click on **Open** button. The **Component Placement** tab will be displayed.
- Set the constrain of recently added assembly to **Default** from **Constrain** drop-down.
- Click on **Done** button from **Component Placement** tab. The **NC Assembly** application window will be displayed along with the added assembly.

Workpiece Wizard

The **Workpiece Wizard** tool is used to assemble workpiece using the Same model, or with features inherited or merged from another model. The procedure to use this tool is discussed next.

- Click on the **Workpiece Wizard** tool of **Components** drop-down from **Manufacturing** tab; refer to Figure-63. The **Open** dialog box will be displayed.

Figure-63. The Workpiece Wizard tool

- Select the specific workpiece for model from **Open** dialog box and click on **Open** button. The **Component Placement** tab will be displayed.
- Click on **Default** from **Constrain** drop-down or you can constrain the workpiece manually and click on **Done** button. The **Create Stock-workpiece** dialog box will be displayed; refer to Figure-64.

Figure-64. The Create Stock-workpiece dialog box for Workpiece Wizard

- Select the **Merge by reference** radio button of **Workpiece Model Type** section, if you want to assemble a workpiece using merge features.
- Select the **Same Model** radio button of **Workpiece Model Type** section, if you want to assemble a workpiece using same model.
- Select the **Inherited** radio button of **Workpiece Model Type** section, if you want to assemble a workpiece using inherited features.
- Click in the **Name** edit box of **Workpiece Model** section and specify the name.
- Click in the **Common name** edit box of **Workpiece Model** section and specify the common name for workpiece.
- After specifying the various parameters click on the **OK** button from **Crate Stock-workpiece** dialog box. The workpiece will be created with the desired features.

Reference Model Wizard

The **Reference Model Wizard** tool is used to assemble a reference model. The procedure to use this tool is discussed next.

- Click on the **Reference Model Wizard** tool of **Component** drop-down from **Manufacturing** tab; refer to Figure-65. The **Open** dialog box will be displayed.

Figure-65. Reference Model Wizard tool

- Select the file to add as reference model and click on **Open** button from **Open** dialog box. The **Component Placement** tab will be displayed.
- Click on **Default** button from **Constrain** drop-down or constrain the reference model manually.
- Click on the **Done** button. The **Create Reference model** dialog box will be displayed; refer to Figure-66.

Figure-66. The Create Reference Model dialog box

- The options of this dialog box were discussed earlier.
- After specifying the various parameters, click on the **OK** button to complete this process. The reference model will be added on **NC Assembly** application window.

PRACTICAL 1

Create a reference model as given in Figure-67 and also create the workpiece for the model.

Figure-67. Practical 1

Starting the NC Assembly

- Double click on **Creo Parametric 4.0** icon from your computer's desktop. The **Creo Parametric 4.0** application window will be displayed; refer to Figure-68.

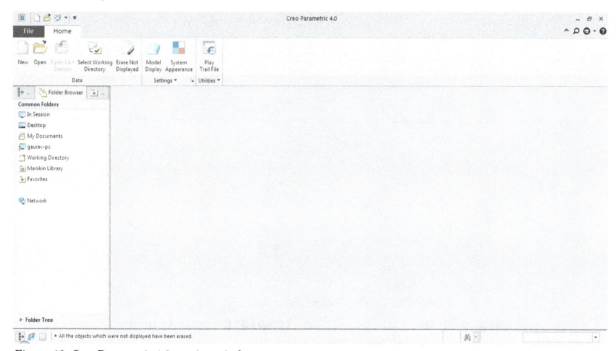

Figure-68. Creo Paramteric 4.0 starting window

- Click on the **New** button from **Ribbon**. The **New** dialog box will be displayed; refer to Figure-69.

Figure-69. New dialog box

- Click on **Manufacturing** radio button from **Type** area and **NC Assembly** radio button from **Sub-Type** area.
- Click on the **Name** edit box and enter **Practical 1** as the name of file.
- Clear the **Use default template** check box from **New** dialog box.
- Click on **OK** button from **New** dialog box. The **New File Options** dialog box will be displayed; refer to Figure-70.

Figure-70. New File Options dialog box

- Click on **mmns mfg nc** option from **Template** section and specify the required details of **Parameters** section.
- Click on **OK** button from **New File Options** dialog box.
- The **NC Assembly** application window will be displayed; refer to Figure-71.

Figure-71. The NC Assembly application window

Adding Reference Model

- Click on **Assemble Reference Model** button of **Reference Model** drop-down from **Manufacturing** tab. The **Open** dialog box will be displayed; refer to Figure-72.

Figure-72. Open dialog box

- Select the **Practical 1** file from resource kit folder and click on **Open** button from **Open** dialog box.

Note- Please download the resource kit folder of Creo parametric 4.0 folder from WWW.CADCAMCAEWORKS.COM.

- The **Component Placement** tab will be displayed along with the model; refer to Figure-73.

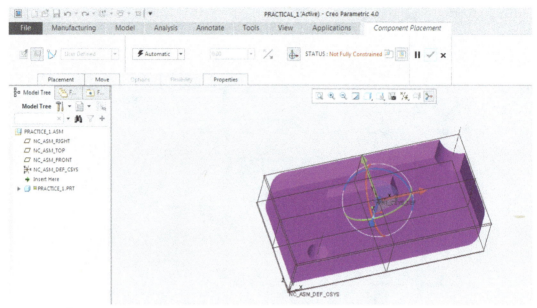

Figure-73. Component Placement tab

- Click on the **Constrain** drop-down from the **Component placement** tab and select **Default** button. The **STATUS** will be changed to **Fully constrained**.
- Click on **Done** button from **Component Placement** tab. The **NC Assembly** application window will be displayed along with the reference model; refer to Figure-74.

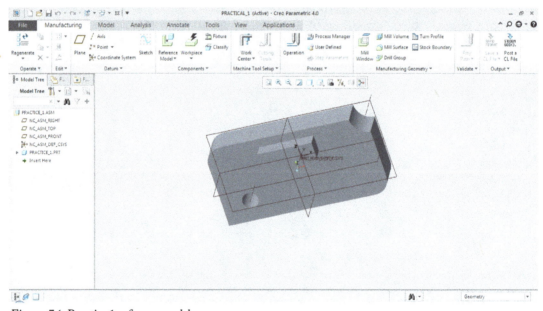

Figure-74. Practice 1 reference model

Creating Workpiece

- Click on the **Automatic Workpiece** button of **Workpiece** drop-down from **Manufacturing** tab. The **Auto Workpiece Creation** tab will be displayed; refer to Figure-75.

Figure-75. Auto Workpiece Creation tab for Practice 1

• Click on **Done** button from **Auto Workpiece Creation** tab. The **NC Assembly** workpiece will be added to reference model; refer to Figure-76.

Figure-76. Model along with workpiece

PRACTICAL 2

Create a reference model as given in Figure-77 and also create the workpiece for the model.

Figure-77. Practical 2

Start the Creo Parametric 4.0 and create a manufacturing part.

Adding Reference Model

- Click on the **Assemble Reference Model** button of **Reference Model** drop-down of **Manufacturing** tab. The **Open** dialog box will be displayed; refer to Figure-78.

Figure-78. Open dialog box

- Select the **Practical 2** file and click on **Open** button from **Open** dialog box. The **Component Placement** tab will be displayed along with the reference model; refer to Figure-79.

Figure-79. Component Placement tab along with Practicle 2 model

- Click on the **Constrain** drop-down of **Component Placement** tab and select **Default** option. The **STATUS** will be changed to **Fully Constrained**.
- Click on **Done** button from **Component Placement** tab. The **NC Assembly** application window will be displayed along with the reference model; refer to Figure-80.

Figure-80. NC Assembly application window after adding model

Adding Workpiece

- Click on **Assemble Workpiece button** of **Workpiece** drop-down from **Manufacturing** tab. The **Open** dialog box will be displayed.
- Select the **Workpiece for Practical 2** file and click on **Open** button. The **Component Placement** tab will be displayed along with the model and workpiece; refer to Figure-81.

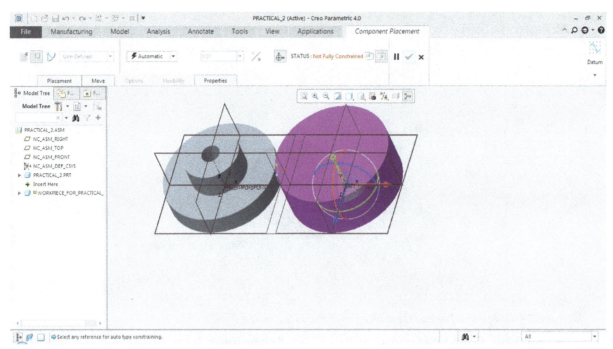

Figure-81. The Component placement tab along with model and workpiece

- Select **Default** option of **constrain** drop-down from **Component Placement** tab Figure-82.

Figure-82. Default option from constrain drop

- On selecting the **Default** button, the reference model will be fully constrained; refer to Figure-83.

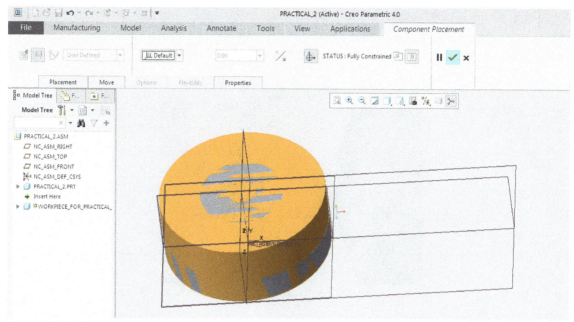

Figure-83. Fully constrained model

• Click on the **Done** button from Component Placement tab. The **NC Assembly** application window will be displayed along with the model; refer to Figure-84.

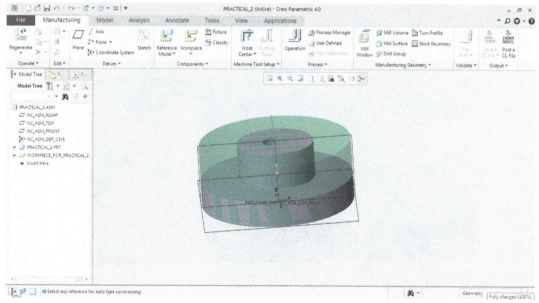

Figure-84. NC Assembly application window for Practical 2

PRACTICE 1

Create the reference for the model shown in Figure-85 and also create the workpiece of model. Part files are available in resource kit.

Figure-85. Practice 1

PRACTICE 2

Create the reference for the model shown in Figure-86 and also create the workpiece of model. Part files are available in resource kit.

Figure-86. Practice 2

Chapter 10

Machine Setup

Topics Covered

The major topics covered in this chapter are:

- *Mill*
- *Mill -Turn*
- *Lathe*
- *Wire EDM*
- *Process Manager*
- *Mill Window*
- *Turn Profile*
- *Stock Boundary*
- *Practice and Practical*

CREATING MACHINE TOOL

By creating machine tool, we are able to work upon workpiece to create a part. In this chapter, we will learn about the basics of creating machine tool and procedure to create tools of Mill, Lathe, Mill-Turn and Wire EDM machine.

WORK CENTER

The **Work Center** tool is used to create different type of machines like Milling machine, Lathe machine, Mill-Turn machine and Wire EDM machine. The procedure to create these machine is discussed next.

Mill

The milling machine is used to remove material from workpiece by using rotary motion of the tool. The procedure to create a milling machine is discussed next.

*   Click on the **Mill** tool of **Work Center** drop-down from **Ribbon**; refer to Figure-1. The **Milling Work Center** dialog box will be displayed; refer to Figure-2.

Figure-1. Mill machine tool

Figure-2. Milling Work Center dialog box

- Click in the **Name** edit box and specify the name for milling machine.
- Click in the **CNC Control** edit box and enter the controller name.
- Click in the **Post Processor** edit box and specify the default name of post processor.
- Click in the **ID** edit box and set a post-processor ID from the list. The **ID** may be from 1 to 99.
- Click on the **Number of Axes** drop-down and select **3-Axis**, **4-Axis** or **5-Axis** for machine.
- Select the **Enable Probing** check box, if you want to allow CMM step definition within an NC session. To use this option you must have the CMM license.

The other options of **Milling Work Center** dialog box have been discussed earlier in Tool Setting chapter.

Mill-Turn

The procedure to create a **Mill-Turn** machine is discussed next.

- Click on the **Mill-Turn** tool of **Work Center** drop-down from **Ribbon**; refer Figure-3. The **Mill-Turn Work Center** dialog box will be displayed; refer to Figure-4.

Figure-3. Mill-Turn tool

Figure-4. Mill-Turn Work Center dialog box

- Click on the **Number of Heads** drop-down and specify the number of head that enables you to set up the cutting tools associated with a machine tool head like if we select **2** from **Number of Heads** drop-down then in **Tools** tab there will be two **Head** section; refer to Figure-5.

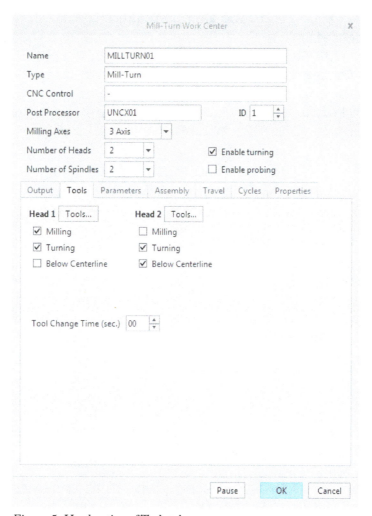

Figure-5. Head section of Tools tab

- Click on the **Number of Spindles** drop-down and select the number of spindle for machine.
- Check the **Enable Turning** check box, if you want to enable turning for machine.
- Check the **Enable Probing** check box if you want to add a probe or default probe from the probe library.

Output tab

There are various options in **Output** tab which are discussed next.

Sub-spindle

- The **Sub-spindle** area is displayed in the dialog box on selecting 2 from the **Number of Spindles** drop-down.
- Select **Main Spindle Program Zero** option from **Program Zero** drop-down, if you want to output all steps using the main spindle program zero.
- Select **Per Spindle Program Zero** option from **Program Zero** drop-down, if you want to output steps with respect to the spindle used to create the steps.

Probe Compensation

- The **Probe Compensation** area is active in the dialog box only if you have selected the **Enable probing** check box.
- Select **Stylus Center** option from **Output Point** drop-down, if you want to select stylus center as the output point.
- Select **Contact Point** option from **Output Point** drop-down, if you want to select contact point of machine as the output point.

Tools tab

There are various tool is this tab which are discussed next.

Head 1

- Select the **Milling** check box if you want to make available milling on a particular head. This is the default selection for **Head 1**.
- Select the **Turning** check box if you want to make available turning on a particular head. This check box is selected by default for all head.
- Select the **Below Centerline** check box, if you want to enable turning below the spindle centerline.
- Click in the **Tool Change Time(sec)** edit box and specify the value of time needed for changing a tool, in seconds. This option is optional.
- Click on the **Tools** button of **Tools** tab from **Mill-Turn Work Center** dialog box; refer to Figure-6. The **Tools Setup (Head 1)** dialog box will be displayed; refer to Figure-7.

Figure-6. Tools tab

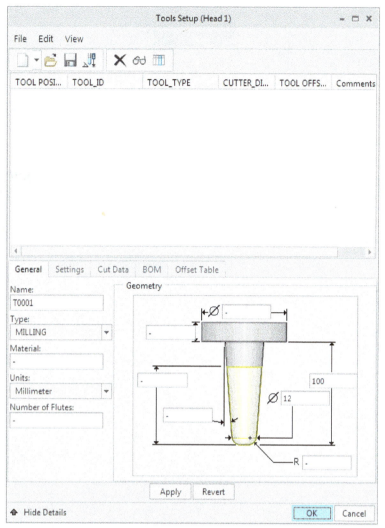

Figure-7. Tool Setup(Head 1) dialog box

The procedure to create a tool was discussed earlier in this book.

- After creating the tool for Head 1, click on **Apply** button. The tool will be created and displayed in the tool list. Note that the tools added in the list will be assumed as installed on the machine as per their tool position in this list.
- Click on **OK** button from **Tool Setup(Head 1)** dialog box.
- The procedure to add or create **Probe** has been discussed earlier in this book.

Parameters tab

- Click in the **Maximum Speed (rpm)** edit box of **Sub-Spindle** section and enter the value of maximum rotating speed of spindle.
- Click in the **Horsepower** edit box of **Sub-Spindle** section and enter the value of horsepower for spindle.
- Click in the **Machine Frequency** edit box of **Sub-spindle** section and enter the value of frequency for machine.
- Click on the **DMIS** button of **Parameters** tab. The **DMIS TEXT Menu Manager** will be displayed; refer to Figure-8. For this option, you must have CMM Probing enabled.

*Figure-8. DMIS
TEXT Menu Manager*

- Click on the **Create** button of **DMIS TEXT Menu Manager**. The **Activate DMIS text** dialog box will be displayed; refer to Figure-9.

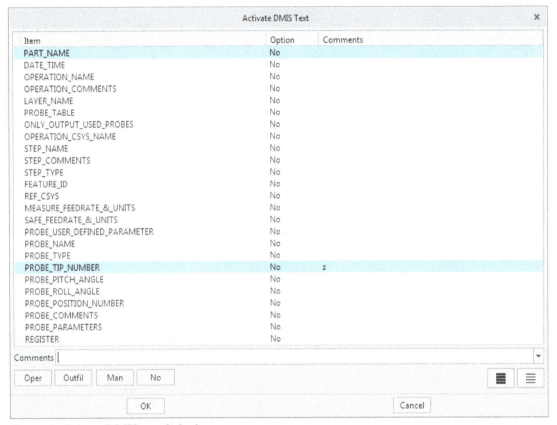

Figure-9. Activate DMIS text dialog box

- Select the specific command from **Item** section and enter the text in **Comments** edit box. The comment will be displayed in Comments section.
- Click on the **Oper** button from **Activate DMIS text** dialog box, if you want to display DMIS Text to CMM screen.
- Click on **Outfil** button from **Activate DMIS text** dialog box, if you want to include DMIS TEXT statement in CMM Output.
- Click on the **Man** button from **Activate DMIS text** dialog box, if you want to display DMIS TEXT statement to CMM operator.
- Click on the **No** button from **Activate DMIS text** dialog box, if you do not want to output or display DMIS TEXT statement.
- After specifying the parameters, click on **OK** button. The **DMIS TEXT** Menu Manager will be displayed again. Click on the **Done/Return** button from Menu Manager. The **Mill-Turn Work Center** dialog box will be displayed.

The other options of this dialog box have been discussed earlier in this book.

Click on **OK** button. The milling machine will be added in Model Tree.

Lathe

- Click on the **Lathe** tool of **Work Center** drop-down from **Ribbon**; refer to Figure-10. The **Lathe Work Center** dialog box will be displayed; refer to Figure-11.

Figure-10. Lathe tool

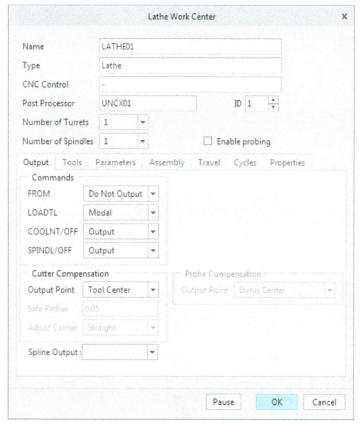

Figure-11. Lathe Work Center dialog box

- Select the number of turret applicable for your machine from the **Number of Turret** drop-down in the **Lathe Work Center** dialog box.
- Select the number of spindle applicable for your machine from the **Number of Spindle** drop-down in **Lathe Work Center** drop-down.

Tools tab

- Click on the **Turret 1** button of **Tools** tab from **Lathe Work Center** dialog box. The **Tools Setup** dialog box will be displayed; refer to Figure-12.

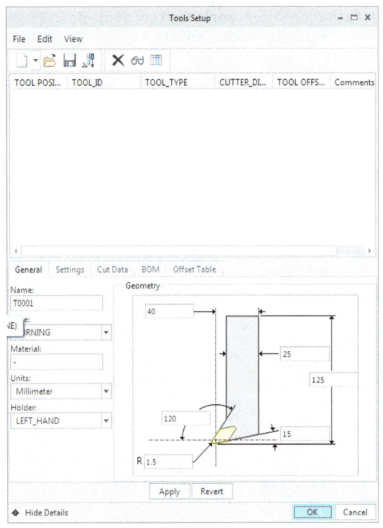

Figure-12. Tools Setup dialog box for turret

- Enter the parameter for tool in specific edit box of geometry section and click on **Apply** button. The tool will be added in Record section.
- After specifying the parameters click on the **OK** button from **Tools Setup** dialog box.
- The Other options of **Lathe Work Center** dialog box are same as discussed earlier in this book. Now, click on the **OK** button from **Lathe Work Center** dialog box to complete this process. The machine will be added in **Model Tree**.

Wire EDM

- Click on the **Wire EDM** tool of **Work Center** drop-down from **Manufacturing** tab; refer to Figure-13. The **WEDM Work Center** dialog box will be displayed; refer to Figure-14.

Figure-13. Wire EDM tool

Figure-14. WEDM Work Center dialog box

Tools tab

- Click on the **Tools** button of **Tools** tab from **WEDM Work Center** dialog box. The **Tools Setup** dialog box will be displayed; refer to Figure-15.

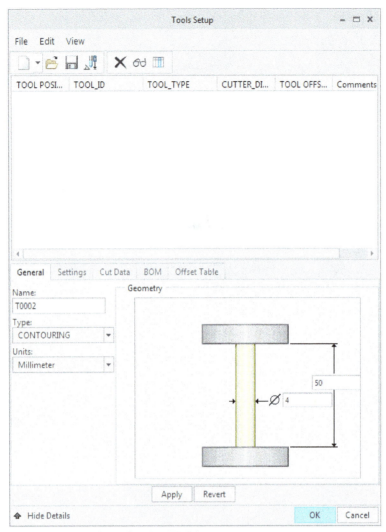

Figure-15. Tools Setup for Wire EDM

- Enter the parameter as required in the specific edit box of **Geometry** section of **General** tab and click on **Apply** button to add the tool in record section.
- Click on the **OK** button of **Tools Setup** dialog box and **WEDM Work Center** dialog box to complete the process. The machine will be added in **Model Tree**.

User-Defined Work Center

The **User-Defined Work Center** tool is used to insert work center from a library of user defined work centers. The procedure to use this tool is discussed next.

- Click on the **User-Defined Work Center** tool of **Work Center** drop-down from **Manufacturing** tab; refer to Figure-16. The **Open** dialog box will be displayed.

Figure-16. User-Defined Work Center tool

• Select the part file from **Open** dialog box and click on **Open** button. The work center will be added in application window.

CREATING AN OPERATION

In any kind of machining, operations are the cutting steps to get the required shape and size from the workpiece. It is possible that a single operation cuts only a small portion of the workpiece and It is also possible that a single operation can cut almost all the workpiece to get the desired parameters in one go.

• Click on the **Operation** tool of **Ribbon** from **Manufacturing** tab; refer to Figure-17. The **Operation** contextual tab will be displayed; refer to Figure-18 and you will be displayed asking you to select a coordinate system.

Figure-17. Operation tool

Figure-18. The Operation tab

• Select the coordinate system to specify orientation of machine and workpiece.

Operation tab

There are various options in **Operation** tab which are discussed next.

Clearance tab

• The options in the **Clearance** tab are used to specify the reference to which the tool will move during retract motion.

- Click on the **Clearance** tab from **Operation** tab; refer to Figure-19.

Figure-19. Clearance tab

- Select **Plane** from **Type** drop-down in **Clearance** tab if you want to define a plane for tool retract.
- Select **Cylinder** from **Type** drop-down in **Clearance** tab if you want to define a plane for tool retract.
- Select **Sphere** from **Type** drop-down in **Clearance** tab if you want to define a plane for tool retract.
- Select **Surface** from **Type** drop-down in **Clearance** tab if you want to define a plane for tool retract.
- After selecting the desired option from the **Type** drop-down, click in the **Reference** selection box and select the desired geometry.
- Click in the **Value** edit box and specify the offset value or radius.
- Click in the **Tolerance** edit box and enter the value of tolerance.
- Clear the **Always use Operation Retract** check box, if you want to use the operation retract only when tool axis orientation changes.

Home Point tab

- Click on the **Home Point** tab from **Operation** contextual tab; refer to Figure-20.

Figure-20. Home Point tab

- Click in the **From** selection box from **Home Point** tab and select the operation starting location.
- Click in **Home** button from **Home Point** tab and select operation end location.

The other options of **Operation** tab have been discussed earlier.

After specifying the parameters, click on the **Done** button from **Operation** tab. The operation will be added in **Modal Tree**. Note that a new tab will be added in the **Ribbon** based on your machine. Like if you have set Milling machine for operations then **Mill** tab will be added.

Process Manager

The **Process Manager** functionality is based on the process table, which lists all the manufacturing process objects, such as workcells, operations, fixture setups, tooling, and NC sequences. When NC sequences are listed in the process table, they are called steps.

- Click on the **Process Manager** tool from **Ribbon**; refer to Figure-21. The **Manufacturing Process Table** dialog box will be displayed; refer to Figure-22.

Figure-21. Process Manger

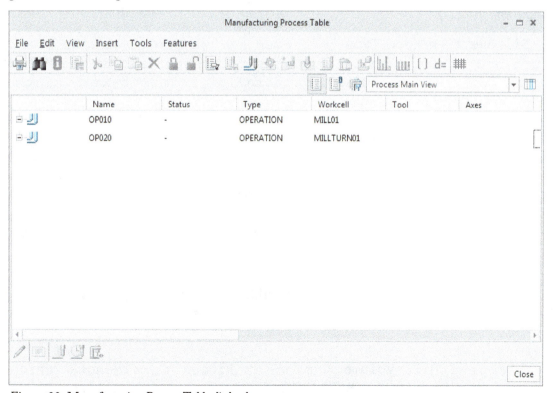

Figure-22. Manufacturing Process Table dialog box

- The list of machine operation will be displayed in record table.
- In **Manufacturing Process Table** dialog box, you can also rename the machine operation name.
- To rename, click on the operation name and enter the desired name.
- Click on **Close** button to close this dialog box.

PREPARING GEOMETRY FOR MANUFACTURING

There are various tools which help us to prepare geometry for manufacturing of part.

Mill Window

The **Mill Window** is used to specify the region in which the tool will move while cutting. The procedure to use this tool is discussed next.

*   Click on the **Mill Window** button of **Ribbon** from **Manufacturing** tab; refer to Figure-23. The **Mill Window** tab will be displayed; refer to Figure-24.

Figure-23. Mill Window button

Figure-24. Mill Window tab

*   The **Silhouette Window Type** button allows you to create the window by projecting the silhouette of the reference part on the **Mill Window** start plane, in the direction parallel to the z-axis of the **Mill Window** coordinate system.
*   The **Sketch Window Type** button allows you to define the **Mill Window** by sketching a closed contour.
*   The **Chain Window Type** button allows you to define the **Mill Window** by selecting edges or curves that form a closed contour.
*   The other options of this tab are same as discussed earlier in this book.
*   Click on the **OK** button from **Mill Window** tab. The mill window will be added in **Model Tree**.

Mill Volume

The **Mill Volume** tool is used to work on volume of a part to be machined. The procedure to use this tool is discussed next.

*   Click on the **Mill Volume** button of **Ribbon** from **Manufacturing** tab; refer to Figure-25. The **Mill Volume** contextual tab will be displayed; refer to Figure-26.

Figure-25. Mill volume tool

Figure-26. Mill Volume tab

Gather Volume Tool

The **Gather Volume Tool** is used to create the volume for milling operations. The procedure to use this tool is discussed next.

• Click on **Gather Volume Tool** button from **Mill Volume** tab. The **VOL GATHER Menu Manager** dialog box will de displayed; refer to Figure-27.

Figure-27. VOL GATHER Menu Manager

• Select the **Select** check box of **GATHER STEPS** section, if you want to select surfaces to be machined.
• Select the **Exclude** check box of **GATHER STEPS** section, if you want to ignore the outer loops or exclude some of the selected surfaces from the volume.
• Select the **Fill** check box of **GATHER STEPS** section, if you want to ignore the inner loops on the selected surfaces.
• Select the **Close** check box of **GATHER STEPS** section, if you want to specify ways of closing the volume, other than the default way described above.

Select the desired check box and click on **Done** button from **Menu Manager**. The updated **Menu Manager** will be displayed; refer to Figure-28.

Figure-28. Updated
Vol GATHRER
Menu Manager

- Select **Surf & Bnd** option from **GATHER SEL** Menu Manager, if you want to select one of the surfaces to be machined (seed surface), and the bounding surfaces.
- Select **Surfaces** option from **GATHER SEL** Menu Manager, if you want to select continuous surfaces to be machined. All the selected surfaces are included in the volume definition.
- Select **Features** option from **GATHER SEL** Menu Manager, if you want to select features to be machined. All the surfaces of selected features are included in the volume definition.
- Select **Mill Surf** option from **GATHER SEL** Menu Manager, if you want to select a pre-defined mill surface from a list.

In our case, we are selecting **Features** option. After selecting the option, click on **Done** button from **GATHER SEL** Menu Manager. The updated Menu Manager will be displayed along with **Select** dialog box; refer to Figure-29.

Figure-29. The VOL
GATHER Menu
Manager

- Select the inner volume of part; refer to Figure-30.

Figure-30. Selection of volume

- After selecting volume, click on the **Done Refs** button from **FEATURE REFS** Menu Manager. The updated Menu Manager will be displayed.

Figure-31. Close Loop Menu Manager

- Select **Cap Plane** check box of **Closure** Menu Manager, if you want to define cap plane of closure.
- Select **All Loops** check box of **Closure** Menu Manager, if you want to select all holes in selected surface to close.
- Select the **Sel Loops** check box of **Closure** Menu Manager, if you want to use bounding loops to close to close the selected volume.

We have selected **Cap Plane** and **Sel Loops** check boxes in our case. Select the plane which closes the selected volume; refer to Figure-32. After selection click on the **Done** button.

Figure-32. Selection of surface

- The Menu Manager will be displayed along with **Select** dialog box. Here you need to select the edges which are closing the selected volume ; refer to Figure-33.

Figure-33. Selection of edges

- After selection of edges, click on **Done/Return** button from the Menu Manager. The gathered volume will be displayed; refer to Figure-34.

Figure-34. Gather volume

- Click on the **OK** button from **Mill Volume** tab to apply or save the changes. The gathered volume will be added in **Model Tree**.

Mill Surface

The **Mill Surface** tool is used to select/create surfaces on which mill machining is to be performed. The procedure to use this tool is discussed next.

- Click on the **Mill Surface** tool of **Manufacturing** tab from **Ribbon**; refer to Figure-35. The **Mill Surface** tab will be displayed; refer to Figure-36.

Figure-35. Mill Surface tool

Figure-36. Mill Surface tab

Most of the tools in tab are the modeling tools. To know more about these tools, please refer to our other book Creo Parametric 4.0 Black Book.

Sloped

The **Sloped Mill Surface** tool is used to easily adjust machining strategy for finish milling based on the orientation of the surface. The procedure to use this tool is discussed next.

- Click on **Sloped** tool from **Mill Surface** tab; refer to Figure-37. The **SURFACE** dialog box will be displayed along with **Select** dialog box; refer to Figure-38.

Figure-37. Sloped tool

Figure-38. SURFACE dialog box

- Here, you need to select the surface for sloped feature. Select the surface as displayed; refer to Figure-39.

Figure-39. Selection of surface for Sloped tool

- Click on **OK** button to select the selection. The **GEN SEL DIR** Menu Manager will be displayed; refer to Figure-40.

Figure-40. GEN SEL DIR Menu Manager

- Select **Plane** option from **GEN SEL DIR** Menu Manager, if you want to select plane as a reference for direction of sloped feature.
- Select **Crv/Edg/Axis** option from **GEN SEL DIR** Menu Manager, if you want to select curve, edge, or axis as a reference for direction of sloped feature.
- Select **Csys** option from **GEN SEL DIR** Menu Manager, if you want to select coordinate system as a reference for direction of sloped feature.

In our case, we are selecting **Crv/Edg/Axis** option. Click on the edge of part as shown in Figure-41. The **DIRECTION** Menu Manager will be displayed.

Figure-41. Selection of edge for slopped feature

Figure-42. DIRECTION
Menu Manager

- Click on **Flip** option of **DIRECTION** Menu Manager, if you want to flip or reverse the selected direction. Otherwise click on **Okay** button. The **Enter draft angle** edit box will de displayed; refer to Figure-43.

Figure-43. Enter draft angle edit box

- Enter the value of draft angle in **Enter draft angle** edit box and click on **Done** button. The **SHALLOW OR STEEP** Menu Manager will be displayed; refer to Figure-44.

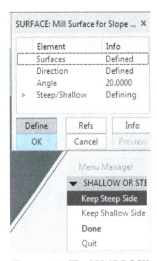

Figure-44. The SHALLOW
OR STEEP Menu Manager

- Select **Keep Steep Side** option from **SHALLOW OR STEEP** Menu Manager, if you want to keep only those portions of the selected surface where the angle between the surface normal and the direction vector is greater than the specified angle value.

- Select **Keep Shallow Side** option from **SHALLOW OR STEEP** Menu Manager, if you want to keep only those portions of the selected surfaces where the angle between the surface normal and the direction vector is less than the specified Angle value.
- In our case we are selecting **Keep Steep Side** option and click on **Done** button. The **SURFACE** dialog box will be displayed.

- Click on **OK** button from **SURFACE** dialog box. The **Sloped** feature will be displayed on application window; refer to Figure-45.

Figure-45. The Slopped feature

Machinable Area Reference

The **Machinable Area Reference** tool create a surface that only includes areas that are to be machined. The procedure to use this tool is discussed next.

- Click on the **Machinable Area Reference** tool from **Mill Surface** tab. The **SURFACE: Machinable Area** dialog box will be displayed along with **Select** dialog box; refer to Figure-46.

Figure-46. The SURFACE dialog box

- Here, you need to select the surface as shown in Figure-47.

Figure-47. Selection of surface for machine

- Click on **OK** button from **Select** dialog box. The Menu Manager will be displayed along with **Select** dialog box; refer to Figure-48.

Figure-48. Menu Manager along with Select dialog box

- Select the surface as reference to check gouging agent from the part as shown in Figure-49.

Figure–49. Selection of surfaces

• Click on **OK** button from **Select** dialog box. The updated Menu Manager will be displayed; refer to Figure-50.

Figure-50. SELECT SRFS
Menu Manager

• Click on **Done/Return** button from all Menu Manager. The **Enter radius of ball end mill for calculating machinable area** edit box will be displayed; refer to Figure-51.

Figure-51. Enter radius of ball end mill for calculating machinable area

- Enter the value in the edit box and click on **Done** button. The **GEN SEL DIR** Menu Manager will be displayed. The options of this dialog box are same as discussed earlier.
- Click on the **Plane** option from **GEN SEL DIR** Menu Manager and select the plane. The **DIRECTION** Menu Manager will be displayed.
- Click on **Okay** button from **DIRECTION** Menu Manager and **OK** from **SURFACE: Machinable Area** dialog box. The surface will be created; refer to Figure-52.

Figure-52. The Surface created after using machinable area reference tool

Other tools or options of Mill Surface tab were discussed earlier in this book.
- After specifying the parameters, click on the **Done** button of **Mill Surface** tab. The commands will be added in **Modal Tree**.

Drill Group

The **Drill Group** tool is used to define a drill group. The procedure to use this tool is discussed next.

- Click on the **Drill Group** tool from **Ribbon** of **Manufacturing** tab; refer to Figure-53. The **Drilling Group** dialog box will be displayed; refer to Figure-54.

Figure-53. Drill Group tool

Figure-54. Drilling Group tool

- Select **Rule diameters** option from **Definition** section, if you want to select holes of specified diameter.
- Select **Rule Surfaces** option from **Definition** section, if you want to select holes of specified surfaces.
- Select **Rule Parameters** option from **Definition** section, if you want to select holes of specified feature parameters.
- Select **Patterned Axes** option from **Definition** section, if you want to select hole in a specific pattern.
- Select **Individual Axes** option from **Definition** section, if you want to select individual holes from drill group.
- Select **Excluded Axes** option from **Definition** section, if you want to exclude holes from drill group.

In our case we are selecting the **Individual Axes** option.

- Select the axis of holes while holding **CTRL** key. The selected axes will be listed in **Individual axes** section; refer to Figure-55.

Figure-55. Selection of axes

- Click on the **OK** button of **Drilling Group** dialog box to apply the changes. The command will be added in **Modal Tree**.

Turn Profile

The **Turn Profile** tool is used to create a cut on part which is represented to turned on lathe machine. The procedure to use this tool is discussed next.

- Click on the **Turn Profile** tool of **Ribbon** from **Manufacturing** tab; refer to Figure-56. The **Turn Profile** tab will be displayed; refer to Figure-57.

Figure-56. Turn Profile tool

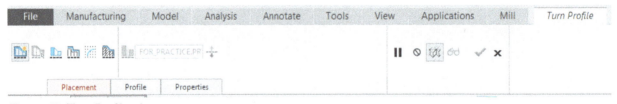

Figure-57. Turn Profile tab

- Click on 🔲 button from **Turn Profile** tab, if you want to define a turn profile by creating an envelope on the fly.
- Click on 🔲 button from **Turn Profile** tab, if you want to select an existing envelope to create a turn profile.

- Click on the button from **Turn Profile** tab, if you want to define turn profile by specifying **From** and **To** surfaces on a reference model.
- Click on the button from **Turn Profile** tab, if you want to define turn profile along the segments of a datum curve or another turn profile.
- Click on the button from **Turn Profile** tab, if you want to sketch a turn profile.
- Click on the button from **Turn Profile** tab, if you want to define a turn profile by section.

Here, we are going to sketch the turn profile so, click on button from **Turn Profile** button tab. Select the coordinate system to be used for machining; refer to Figure-58.

Figure-58. Selecting coordinate system

- Now, click on the button from **Turn Profile** tab. The **Sketch** dialog box will be displayed .
- Click on **Sketch** button from **Sketch** dialog box. The **References** dialog box will be displayed; refer to Figure-59.

Figure-59. References dialog box

- Here, you need to select the reference to fully place the model. Select the coordinate system as reference as shown in Figure-60

Figure-60. Selection of coordinate system as reference

- Click on **Solve** button from **References** dialog box. If **Fully Placed** is displayed in **References** status section, then click on **Close** button. Otherwise fully placed the model by selecting other references.
- The **Sketch** tab will be displayed. Select the **Project** tool of **Ribbon** from **Sketch** tab. The **Type** dialog box will be displayed; refer to Figure-61

Figure-61. Type dialog box

- Select the edges of part as shown in Figure-62 and click on **OK** button from **Sketch** tab. The **Turn Profile** tab will be displayed again.

Figure-62. Selection of sketch

- Click on **Done** button from **Turn Profile** tab. The profile command will be added to **Model Tree**.

Stock Boundary

The **Stock Boundary** tool is used to create the boundary of stock or workpiece. The procedure to use this tool is discussed next.

- Click on the **Stock Boundary** tool of **Ribbon** from **Manufacturing** tab; refer to Figure-63. The **Stock Boundary** contextual tab will be displayed; refer to Figure-64.

Figure-63. Stock Boundary tool

Figure-64. Stock Boundary tab

- Select the coordinate system from part as a reference. The **Stock Boundary** will automatically created; refer to Figure-65.

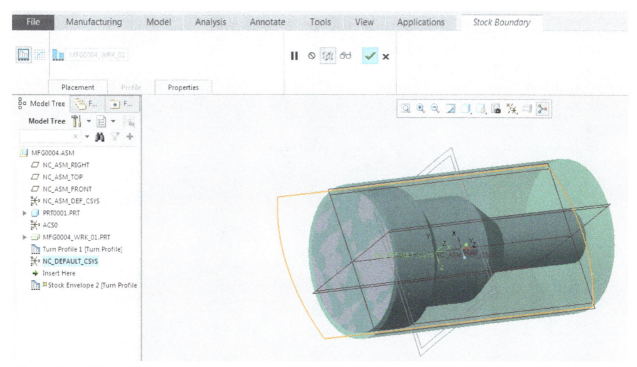

Figure-65. Stock Boundary

- Select the 🔲 button from **Stock Boundary** tab, if you want to create the boundary for workpiece automatically.
- Select the 🔲 button from **Stock Boundary** tab, if you want to create a boundary for workpiece with the help of sketch.

- After specifying the parameters for **Stock Boundary**, click on **Done** button from **Stock Boundary** tab. The boundary will be added in **Model Tree** with the name of **Stock Envelope**.

PRACTICAL 1

Create the milling operation for the model as shown in Figure-66.

Figure-66. Practical 1

Starting the NC Assembly

- Double click on **Creo Parametric 4.0** icon from your computer's desktop. The **Creo Parametric 4.0** application window will be displayed; refer to Figure-67.

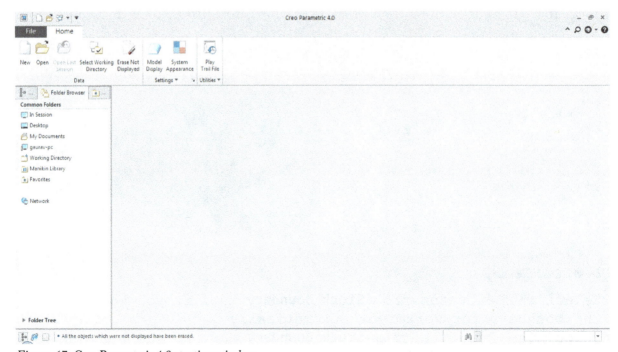

Figure-67. Creo Paramteric 4.0 starting window

- Click on the **New** button from **Ribbon**. The **New** dialog box will be displayed; refer to Figure-68.

Figure-68. New dialog box

- Click on **Manufacturing** radio button from **Type** area and **NC Assembly** radio button from **Sub-Type** area.
- Click on the **Name** edit box and enter **Practical 1** as the name of file.
- Clear the **Use default template** check box from **New** dialog box.
- Click on **OK** button from **New** dialog box. The **New File Options** dialog box will be displayed; refer to Figure-69.

Figure-69. New File Options dialog box

- Click on **mmns mfg nc** option from **Template** section and specify the required details of **Parameters** section.
- Click on **OK** button from **New File Options** dialog box.

• The **NC Assembly** application window will be displayed; refer to Figure-70.

Figure-70. The NC Assembly application window

Adding Reference Model

• Click on **Assemble Reference Model** button of **Reference Model** drop-down from **Manufacturing** tab. The **Open** dialog box will be displayed; refer to Figure-71.

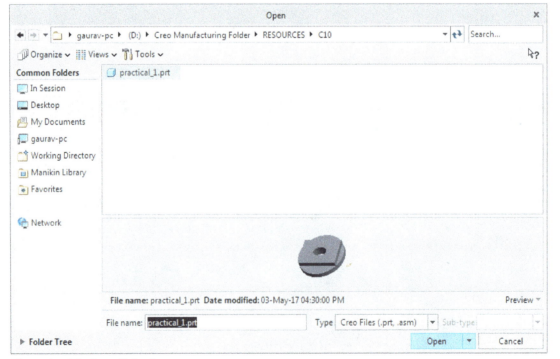

Figure-71. The Open dialog box

• Select the **Practical 1** file and click on **Open** button from **Open** dialog box. The **Component Placement** tab will be displayed.

- Click on **Constrain** drop-down and select the **Default** button. The **STATUS: Fully Constrained** will be displayed in **Component Placement** tab; refer to Figure-72.

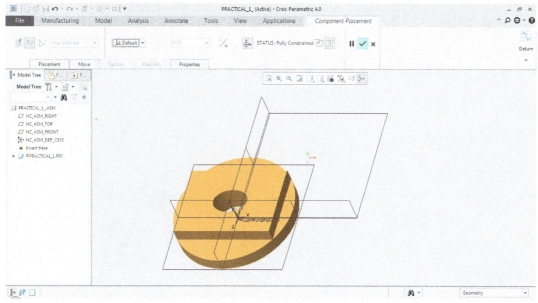

Figure-72. Fully constrained model

- Click on **Done** button from **Component Placement** tab. The **NC Assembly** application window will be displayed along with the constrained model.

Crating Workpiece

- Click on the **Automatic Workpiece** button of **Workpiece** drop-down from **Manufacturing** tab. The **Auto Workpiece creation** tab will be displayed; refer to Figure-73

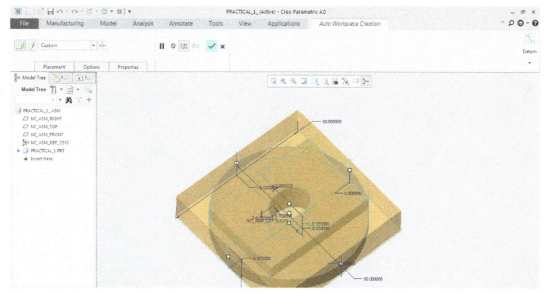

Figure-73. Creating workpiece for model

- Click on **Create Round Workpiece** button from **Auto Workpiece Creation** tab.
- Click on **Options** tab of **Auto Workpiece Creation** tab.
- Click on **About X** edit box of **Rotation Offset** section and enter the value as **90**; refer to Figure-74.

- Click on **Length(+)** button of **Options** tab and enter the value as **0**.
- Click on **Length(-)** button of **Options** tab and enter the value as **0**.

Figure-74. Options tab

- After specifying the parameters, click on **Done** button from **Auto Workpiece Creation** tab.
- The **NC Assembly** application window will be displayed along with the reference model and workpiece.

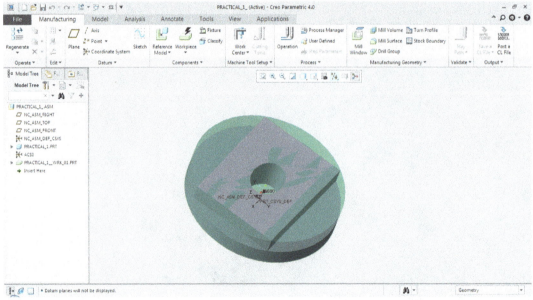

Figure-75. NC Assembly application window along with model and woorkpiece

Machine Tool Setup

- Click on the **Mill** tool from **Work Center** drop-down of **Manufacturing** tab. The **Milling Work Center** dialog box will be displayed Figure-76.

Figure-76. Milling Work Center dialog box

- Click on **Name** edit box and enter **TECHNO** as the name of machine.
- Click on **Tools** button of **Tools** tab from **Milling Work Center** dialog box. The **Tools Setup** dialog box will be displayed; refer to Figure-77.
- Click on **Type** drop-down and select **END MILL** button.
- Specify the parameters of the tool as displayed in figure and click on **Apply** button. The tool will be added in record section.
- Similarly create an another tool of parameters displayed in Figure-78

Figure-77. Tools Setup dialog box for practical 1

Figure-78. Creting another tool

- Click on **Settings** tab of **Tools Setup** dialog box and enter 2 in **Tool Number** edit box.

- Click on **Apply** button and the tool will be added in records section of **Tools Setup** dialog box.
- After specifying the parameters, click on **OK** button from **Tools Setup** dialog box. The **Milling Work Center** dialog box will be displayed.
- Click on **Parameters** tab of **Milling Work Center** dialog box and enter 6000 in **Maximum Speed(RPM)** edit box.
- Similarly, specify the other parameters, click on **OK** button from **Milling Work Center** dialog box. The machine will be added in Model Tree with the name as **TECHNO**.

Creating Operation

- Click on the **Operation** button of **Process** panel from **Manufacturing** tab. The **Operation** tab will be displayed; refer to Figure-79.

Figure-79. The Operation tab

- Click on **Datum** button from **Ribbon** and select **Coordinate System** button; refer to Figure-80. The **Coordinate System** dialog box will be displayed; refer to Figure-81.

Figure-80. Datum

Figure-81. Coordinate System dialog box

- Select the surface of model as shown in Figure-82. The coordinate system will displayed on selected surface.

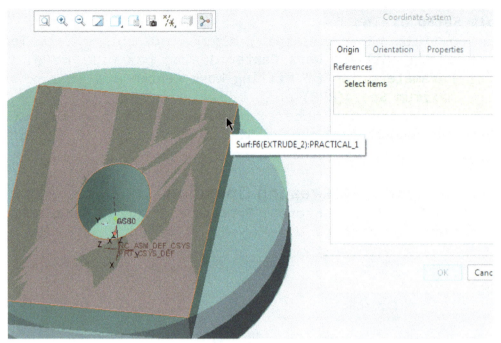

Figure-82. Selection of surface for Coordinate System

- Now, drag the green button of coordinate system to the edge of model to define reference; refer to Figure-83.

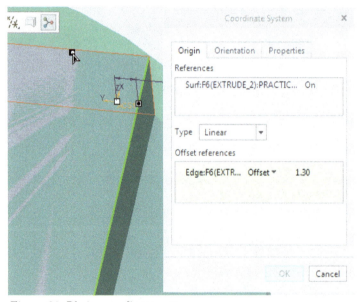

Figure-83. Placing coordinate system

- Click on the **Offset** drop-down of **Offset** references section from **Coordinate System** dialog box and select the **Align** option in both the drop-down; refer to Figure-84.
- Click on **Orientation** tab of **Coordinate system** dialog box and select **Y** from to **project** drop-down.
- Click on **Flip** button of **to Project** section. The direction of coordinate system will be changed in such a way that Z axis is perpendicular to the milling face of model; refer to Figure-85.

Figure-84. Aligning the coordinate system

Figure-85. Change of direction of coordinate system

- Click on **OK** button from **Coordinate System** dialog box. The **Operation** tab will be displayed.
- Click on **Resume** button from **Operation** tab and the newly created coordinate system will be selected automatically.
- Click on **Done** button from **Operation** tab. The operation command will be added in Model Tree with the name as **OPO10[TECHNO]**.

Creating Mill Window

- Click on **Mill Window** button of **Manufacturing** Geometry panel from **Manufacturing** tab. The **Mill Window** tab will be displayed; refer to Figure-86.

Figure-86. Mill Window tab

- Click on **Window Plane** selection box of **Placement** tab and click on upper face of workpiece; refer to Figure-87.

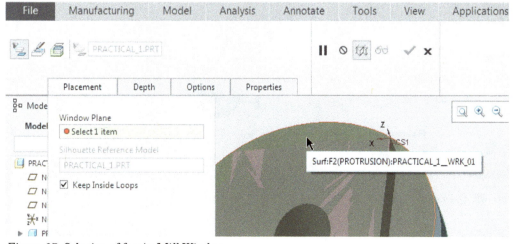

Figure-87. Selection of face in Mill Window

- Click on **Done** button from **Mill Window** tab. The **NC Assembly** application window will be displayed.

PRACTICAL 2
Apply the Lathe operation in Figure-88.

Figure-88. Practical 21

Start the NC Assembly application as discussed earlier.

Adding Reference Model

- Click on **Assemble Reference Model** button of **Reference Model** drop-down from **Manufacturing** tab. The **Open** dialog box will be displayed; refer to Figure-89.

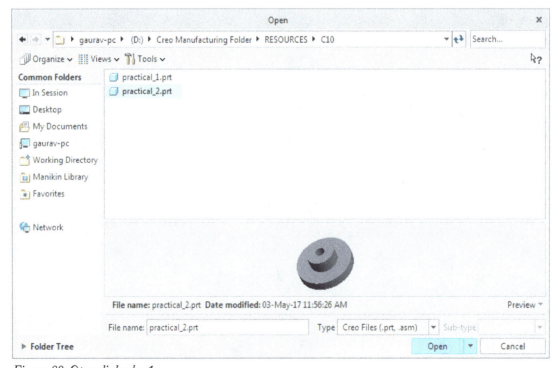

Figure-89. Open dialog box1

- Select the **Practical 2** file and click on **Open** button from **Open** dialog box. The **Component Placement** tab will be displayed.
- Click on **Constrain** drop-down and select the **Default** button. The **STATUS: Fully Constrained** will be displayed in **Component Placement** tab; refer to Figure-90.

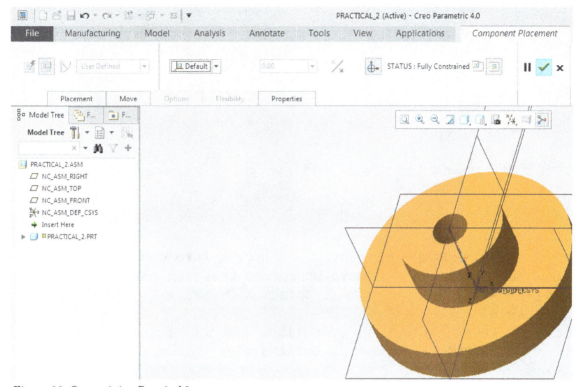

Figure-90. Constraining Practical 2

- Click on **Done** button from **Component Placement** tab. The **NC Assembly** application window will be displayed along with the constrained model.

Crating Workpiece

- Click on the **Automatic Workpiece** button of **Workpiece** drop-down from **Manufacturing** tab. The **Auto Workpiece creation** tab will be displayed; refer to Figure-91.

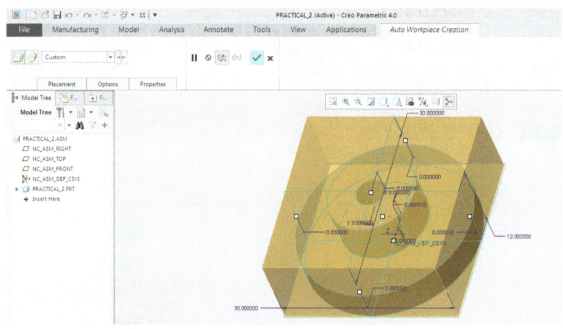

Figure-91. Creating workpiece for practical 2 model

- Click on **Create Round Workpiece** button from **Auto Workpiece Creation** tab.
- Click on **Options** tab of **Auto Workpiece Creation** tab.
- Click on **About X** edit box of **Rotation Offset** section and enter the value as **90**; refer to Figure-92.
- Click on **Length(+)** button of **Options** tab and enter the value as **0**.
- Click on **Length(-)** button of **Options** tab and enter the value as **0**.

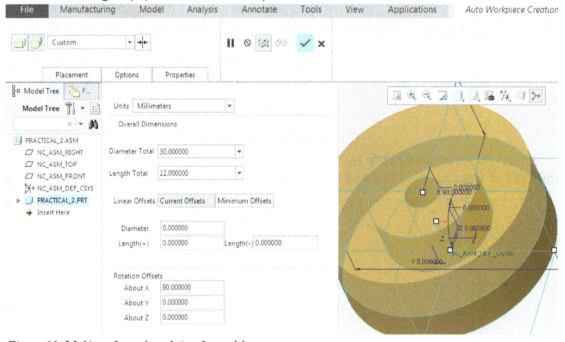

Figure-92. Making of round workpiece for model

- Click on **Done** button from **Auto Workpiece Creation** tab. The **NC Assembly** application window will be displayed along with model and workpiece.

Creating Lathe Operation

- Click on **Lathe** tool of **Work Center** dialog box from **Manufacturing** tab. The **Lathe Work Center** dialog box will be displayed; refer to Figure-93.

Figure-93. Lathe Work Center dialog box

- Click in the **Name** edit box and enter **TMT LATHE** as the name of machine.
- Click on **Tools** tab of **Lathe Milling Center** dialog box and click on **Turret 1** button of **Tools** tab. The **Tools Setup** dialog box will be displayed.
- Enter the parameters of tool as shown in Figure-94.

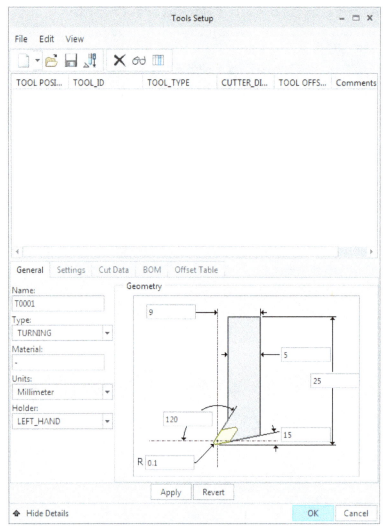

Figure-94. Tools Setup dialog box for lathej

- Click on **Apply** button from **Tools Setup** dialog box. The tool will be added in the list.
- Click on **OK** button from **Tools Setup** dialog box and **Lathe Work Center** dialog box. The **NC Assembly** application window will be displayed; refer to Figure-95.

Figure-95. Model after adding machine

Creating Operation

• Click on the **Operation** button of **Process** panel from **Manufacturing** tab. The **Operation** tab will be displayed; refer to Figure-96.

Figure-96. Operation tab

• Click on coordinate system from model to set the program zero point and click on **Done** button from **Operation** tab.

Creating Turn Profile

• Click on **Turn Profile** tool of **Manufacturing Geometry** panel from **Manufacturing** tab. The **Turn Profile** tab will be displayed; refer to Figure-97.

Figure-97. Turn Profile tab

• Click on the **Placement** tab of **Turn Profile** tab and select the coordinate system.
• Make sure the Z direction of your coordinate system is perpendicular to the flat face of model; refer to Figure-98.

Figure-98. Selection of coordinate system

- Click on the ⊞ button from **Turn Profile** tab. The **Sketch** dialog box will be displayed. Click on **Reference** selection box and select the reference for sketch as shown in Figure-99.

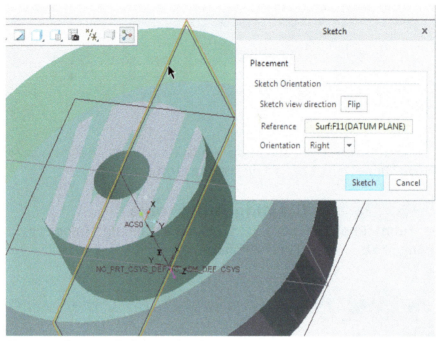

Figure 99. Selection of reference in sketch dialog box

- Click on **Sketch** button from **Sketch** dialog box. The **Sketch** window will be displayed along with **References** dialog box; refer to Figure-100.

Figure-100. References dialog box

- Select the coordinate system from window and click on **Solve** button of **References** dialog box.
- If **Fully Placed** message is shown in **References status** dialog box, then click on **Close** button. The **Sketch** window will be displayed; refer to Figure-101.

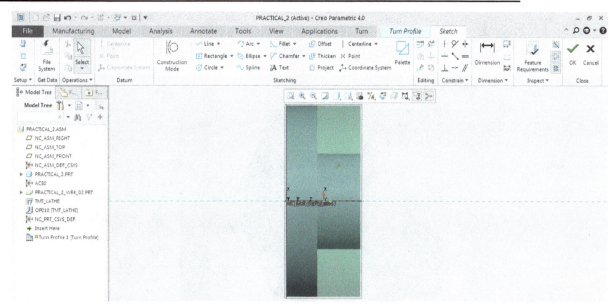

Figure-101. Sketch window

- Left click on the highlighted point and hold the right click of mouse. The **Menu** will be displayed; refer to Figure-102.

Figure-102. Menu

- Click on the **Start Point** from the **Menu** displayed to set the selected point as starting point of machine.
- Click **OK** button from **Sketch** tab. The **Turn Profile** tab will be displayed.
- Click on **Done** button from **Turn Profile** tab. The Turn Profile operation will be added in Model Tree.

PRACTICE 1

Crete a tool in Lathe machine and also create a turn profile; refer to Figure-103.

Figure-103. Practice 1

PRACTICE 2

Crete a tool in Mill machine and also create Mill Window; refer to Figure-104

Figure-104. Practice 2

FOR STUDENTS NOTES

FOR STUDENTS NOTES

Chapter 11

Milling Operations

Topics Covered

The major topics covered in this chapter are:

- *Milling Operations*
- *Profile Milling*
- *Surface Milling*
- *2-Axis Trajectory*
- *Standard Drilling Tool*
- *Deep Drilling Tool*
- *Practice and Practical*

INTRODUCTION

In previous chapters, we have learned the procedure to setting up a machine for particular operation. In this chapter, we will learn about various operations and tools of mill machine.

SETUP A MILLING OPERATIONS

The procedure to setup a milling operation is discussed next.

- Click on the **Operation** tool of **Process** panel from **Manufacturing** tab; refer to Figure-1. The **Operation** tab will be displayed; refer to Figure-2.

Figure-1. Operation tool

Figure-2. The Operation tab

- Select the coordinate system as a program zero point and click on **Done** button from Operation tab. The **Mill** tab will be displayed after **Applications** tab; refer to Figure-3.

Figure-3. Mill tab

Note- The **Mill** tab is displayed when you have selected a milling machine to perform operations.

MILLING OPERATION

There are various milling operation in Creo Parametric 4.0 like Face Milling, Surface Milling, Profile Milling etc. These operations are discussed next.

Roughing Toolpath

The **Roughing** is used to remove unwanted material from workpiece. The procedure to use this tool is discussed next.

- Click on the **Roughing** tool of **Roughing** drop-down from **Mill** tab; refer to Figure-4. The **Roughing** contextual tab will be displayed; refer to Figure-5

Figure-4. Roughing tool

Figure-5. Roughing Tab

- Select the specific tool for roughing from **Tools** drop-down which you have created earlier.
- Click on the **Reference** tab of **Roughing** tab. The earlier created **Mill Window** for workpiece will be automatically selected; refer to Figure-6. If not selected automatically, then select it.

Figure-6. Reference tab

- Now, click on the **Parameters** tab of **Roughing** contextual tab and specify the parameters for the roughing operations; refer to Figure-7.

Figure-7. Parameters tab of Roughing operation

- Click on the **Clearance** tab of **Roughing** contextual tab; refer to Figure-8.

Figure-8. Tthe Clearance tab

- Click in the **Reference** selection box from **Retract** section and select the reference for cutting tool.
- After specifying the parameters, click on the **Done** button from **Roughing** tab. The **Roughing** toolpath will be added in Modal Tree.

Volume Rough

The **Volume Rough** tool is used to rough large amount of stock material. The procedure to use this tool is discussed next.

- Click on the **Volume Rough** tool of **Roughing** drop-down from **Mill** tab; refer to Figure-9. The **Volume Milling** contextual tab will be displayed; refer to Figure-10.

Figure-9. Volume Rough tool

Figure-10. The Volume Milling tab

- Click on the **Reference** tab from **Volume Milling** contextual tab; refer to Figure-11.

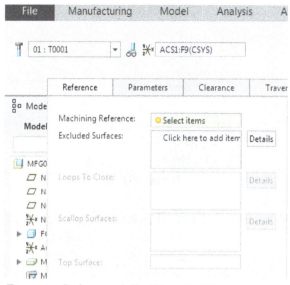

Figure-11. Reference tab For Volume Milling

- Click in the **Machining Reference** selection box of **Volume Milling** tab and select the top surface of workpiece.
- Click on the **Parameters** tab of **Volume Milling** contextual tab and specify the necessary parameters.
- Click on the ▣ button from **Volume Milling** tab, if you want to see the tool path for part.
- After specifying the parameters, click on the **Done** button. The **Volume Milling** toolpath will be added in **Modal Tree**.

Face

The **Face** tool is used to apply a face milling operation on the workpiece. The procedure to use this tool is discussed next.

- Click on the **Face** tool from **Mill** tab; refer to Figure-12. The **Face Milling** contextual tab will be displayed; refer to Figure-13.

Figure-12. Face Tool

Figure-13. Face milling tab

- Click on the **Reference** tab of **Face Milling** contextual tab; refer to Figure-14.

Figure-14. Reference tab for Fce ,Milling

- Select **Surface** option from **Type** drop-down, if you want to select the surface as a reference.
- Select **Mill Window** option from **Type** drop-down, if you want to select the mill window as a machining reference.
- In our case, we are selecting **Mill Window**. Click on the **Machining References** selection box and select the earlier created mill window from **Model Tree**; refer to Figure-15.

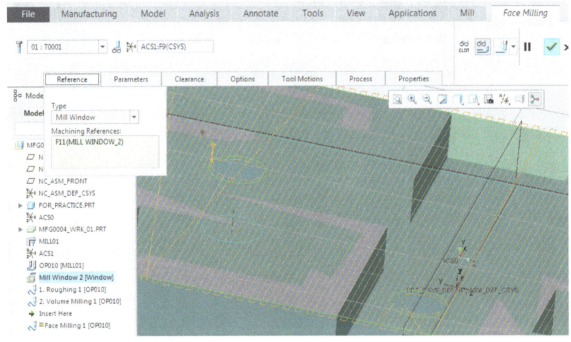

Figure-15. Selection of mill window

- Similarly you can use the **Surface** option of **Type** drop-down.
- Click on the **Parameters** tab from **Face Milling** tab and specify the necessary parameters.
- After specifying the parameters, click on **Done** button from **Face Milling** tab. The **Face Milling** operation will be added in **Model Tree**.

Re-Rough tool

The **Re-Rough** tool is used to re-rough the workpiece with finer tool. The procedure to use this tool is discussed next.

- Click on the **Re-Rough** tool from **Mill** tab; refer to Figure-16. The **Re-roughing** contextual tab will be displayed; refer to Figure-17.

Figure-16. Re-Rough tool

Figure-17. Re-Roughing Tab

- Select the machining tool from **Tools Selection** drop-down.
- Select the roughing operation earlier performed from the previous Operation drop-down in the **Re-roughing** contextual tab of Ribbon..
- Click on the **Parameters** tab from **Re-roughing** contextual tab and specify the necessary parameters for machining process.
- After specifying the parameters click on the **Done** button. The re-roughing operation will be added in **Modal tree**.

Profile Milling

Profile milling is used to remove material from outer walls of the workpiece. The procedure to use **Profile Milling** tool is discussed next.

- Click on the **Profile Milling** tool from **Mill** tab; refer to Figure-18. The **Profile Milling** tab will be displayed; refer to Figure-19.

Figure-18. Profile Milling tool

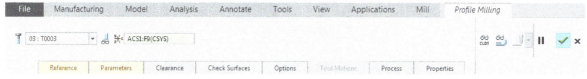

Figure-19. Profile milling tab

- Select the tool from **Tool Selection** drop-down.
- Click on the **References** tab from **Profile Milling** tab; refer to Figure-20.

Figure-20. References tab of Profile Milling

- Select the **Surface** option from **Type** drop-down.
- Click on the **Machining References** section and select the surface of workpiece; refer to Figure-21.

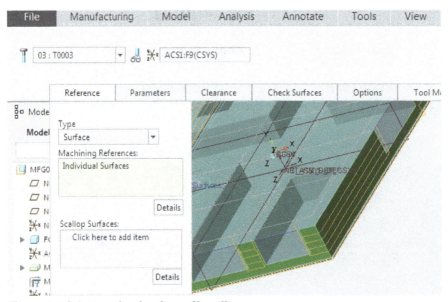

Figure-21. Selection of surface for profile milling

- Click on the **Parameters** tab and specify the necessary parameters for profile milling.
- After specifying the parameters, click on the **Done** button from **Profile Milling** tab. The profile milling operation will be added in **Model Tree**.

Surface Milling

The **Surface Milling** is usually used to semi-finish or finish the surfaces. The procedure to use this tool is discussed next.

- Click on the **Surface Milling** tool from **Mill** tab; refer to Figure-22. The **Menu Manager** will be displayed; refer to Figure-23.

Figure-22. Surface Milling tool

- Select the **Name** check box from **SEQ SETUP** Menu Manager, if you want to specify the name of surface milling operation. Similarly select the required check box from **SEQ SETUP** Menu Manager.
- After selecting the required check box, click on **Done** button from **SEQ SETUP** Menu Manager. The **Tools Setup** dialog box will be displayed if you have not selected the **Name** and **Comments** check boxes in the Menu Manager; refer to Figure-24.

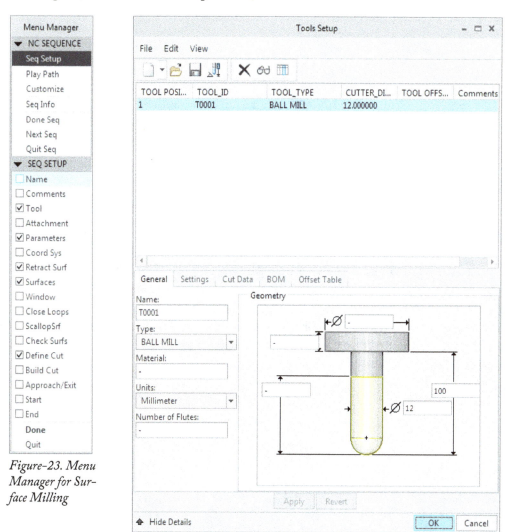

Figure-23. Menu Manager for Surface Milling

Figure-24. Tools Setup dialog box

- Create the tool if not created yet and click on **OK** button from **Tools Setup** dialog box. The **Edit Parameters of Sequence** dialog box will be displayed; refer to Figure-25.

Figure-25. Edit Parameters of sequence dialog box

• Here you need to specify the parameters for the surface milling operations.
• After specifying the necessary parameters for surface milling operation, click on the **OK** button from **Edit Parameters of Sequence** dialog box. The **Retract Setup** dialog box will be displayed; refer to Figure-26.

Figure-26. Retract Setup dialog box

• Click on the **value** edit box and enter the retract value.
• Click on the **OK** button from **Retract Setup** dialog box. The **SURF PICK** Menu Manager will be displayed; refer to Figure-27.

Figure-27. SURF PICK Menu Manager

- Click on the **Done** button from **SURF PICK** Menu Manager. The **Select** dialog box will be displayed.
- Select the surface from workpiece and click on the **OK** button from **Select** dialog box.
- Click on the **Done/Return** button from **SELECT SRFS** Menu Manage. The **Cut Definition** dialog box will be displayed; refer to Figure-28.

Figure-28. Cut Definition dialog box

- Select **Straight cut** radio button from **Cut Type** section, if you want to machine the surface in straight line.
- Select **From Surface Isolines** radio button from **Cut Type** section, if you want to machine the surface along lines that define the surface.
- Select **Projected Cut** radio button from **Cut Type** section, if you want to machine the surface along the selected curve or edge.
- After specifying the parameters, click on the **OK** button from **Cut Definition** dialog box.

- Click on the **Done Seq** button from **NC SEQUENCE** Menu Manager; refer to Figure-29. The surface milling operation will be added in **Modal Tree.**

*Figure-29. NC
SEQUENCE
Menu Manager*

Finishing

The **Finishing** tool is use to give the final touch to the workpiece. The procedure to use this tool is discussed next.

- Click on the **Finishing** tool from **Mill** tab; refer to Figure-30. The **Finishing** contextual tab will be displayed; refer to Figure-31.

Figure-30. Finishing tool

Figure-31. Finishing Tab

- Click on **tools** selection box and select the required tool.
- Click on **References** tab and select the **Mill Window.**
- Click on **Parameters** tab and specify the required parameters for machining.
- Click on **Done** button from **Finishing** tab. The **Finishing** operation will be added in **Model Tree;** refer to Figure-32.

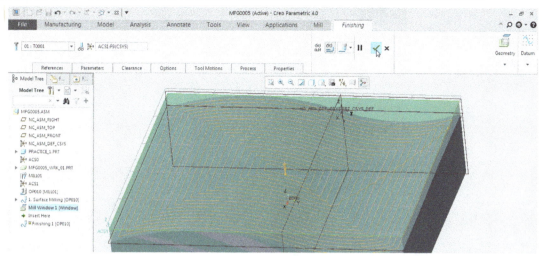

Figure-32. Preview of toolpath

Corner Finishing

The **Corner Finishing** tool is used on corners of workpiece. The procedure to use this tool is discussed next.

- Click on the **Corner Finishing** tool from **Mill** tab; refer to Figure-33. The **Corner Finishing** tab will be displayed; refer to Figure-34.

Figure-33. Corner Finishing tool

Figure-34. Corner Finishing tab

- Click on the **Tool Selection** drop-down and select the required tool for machining.
- Click on the **References** tab of **Corner Finishing** contextual tab and select the required tool in **Reference Cutting Tool** drop-down; refer to Figure-35.

Figure-35. References tab of corner finishing

- Click on the **Parameters** tab and specify the required parameters for machining.
- Click on the **Clearance** tab and select the retract plane using **Reference** selection box of **Retract** section; refer to Figure-36.

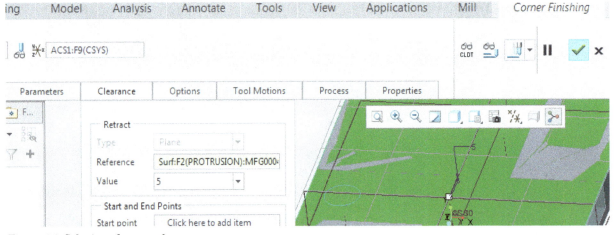

Figure-36. Selection of retract plane

- After specifying the parameters for machining, click on the **Done** button from **Corner Finishing** tab. The **Corner Finishing** operation will be added in **Model Tree**.

2-Axis Trajectory

The **2-Axis Trajectory** tool is used to remove material by following a selected trajectory. The procedure to use this tool is discussed next.

- Click on the **2-Axis Trajectory** tool from **Trajectory Milling** drop-down of **Mill** tab; refer to Figure-37.The **Curve Trajectory** contextual tab will be displayed; refer to Figure-38.

Figure-37. 2-Axis Trajectory

Figure-38. Curve Trajectory tab

- Click on the **Tool Selection** drop-down and select the required tool for machining.
- Click on the **References** tab of **Curve Trajectory contextual** tab. Click on the **Machining References** selection box and select the reference for trajectory; refer to Figure-39. To select multiple references, click on the Details button and select the references while holding the **CTRL** key.

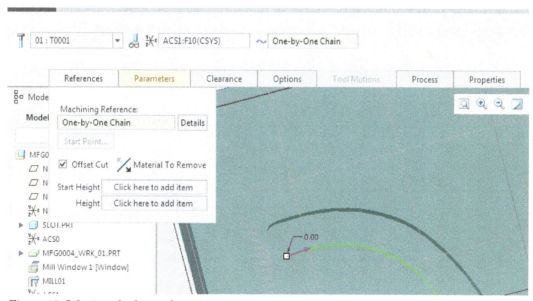

Figure-39. Selection of reference for trajectory

- Click on **Parameters** tab and specify the required parameters for machining.
- After specifying the parameters, click on **Done** button from **Curve Trajectory** tab. The curve trajectory operation will be added in **Model Tree**.

Trajectory

The procedure to use **Trajectory** tool is discussed next.

- Click on the **Trajectory** tool from **Trajectory Milling** drop-down; refer to Figure-40. The **Trajectory** tab will be displayed; refer to Figure-41.

Figure-40. Trajectory tool

Figure-41. Trajectory tab

- Select the required tool for machining from tools selection box.
- Click on **Parameters** tab and specify the required parameters for machining.
- Click on **Tool Motions** tab and click on **Curve Cut** button. The **Curve Cut** dialog box will be displayed; refer to Figure-42.

Figure-42. Curve Cut dialog box

- Click in the **Trajectory Curve** selection box and select the trajectory for machining.
- Click on **OK** button from the **Curve Cut** dialog box and **Tool Motions** tab will be displayed; refer to Figure-43.

Figure-43. The Tool Motions tab

- Click on **Done** button from **Trajectory** tab. The trajectory operation will be added in Model Tree.

Engraving

The **Engraving** tool is used to machine grooves or curves on the metal surface. This can also be used to machine letters on the surface of metal. The procedure to use this tool is discussed next.

- Click on the **Engraving** tool from Mill tab; refer to Figure-44. The **Engraving** tab will be displayed; refer to Figure-45.

Figure-44. Engraving tool

Figure-45. Engraving tab

- Click on the **Tool Selection** drop-down and select the required tool for machining.
- Click on the **References** tab and select the groove to be machined; refer to Figure-46.

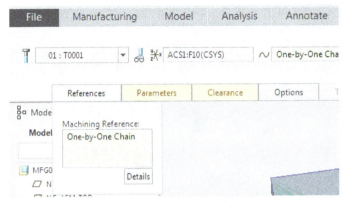

Figure-46. References tab for engraving tool

- Click on the **Parameters** tab and specify the required parameters for machining.
- Click on the **Clearance** tab and select the retract plane in **References** section; refer to Figure-47.

Figure-47. Selection of retract plane for engraving

- Click in the **Value** edit box of **Retract** section and specify the value for retract.
- After specifying the parameters, click on the **Done** button. The **Engraving operation** will be added in **Model Tree**.

Standard

The **Standard** tool is used to drill holes in the workpiece. The procedure to use this tool is discussed next.

- Click on the **Standard** tool of **Holemaking Cycles** panel from **Mill** tab; refer to Figure-48. The **Drilling** tab will be displayed; refer to Figure-49.

Figure-48. Standard tool

Figure-49. Drilling tab

- Click on the **Tool Selection** drop-down from **Mill** tab and select the required tool for drilling.
- Click on **References** tab of **Drilling** contextual tab; refer to Figure-50.

Figure-50. Parameters tab for drilling operation

- Select **Axes** option from **Type** drop-down, if you want to select axes as references for drilling.
- Select **Points** option from **Type** drop-down, if you want to select points as references for drilling.
- Select **Geometry** option from **Type** drop-down, if you want to select geometry as reference for drilling.
- In our case, we are selecting **Axes** option from **Type** drop-down.
- Click in the **Holes** selection box and select the axes of hole for drilling.
- Click on the **Parameters** tab and specify the required parameters for drilling.
- After specifying the parameters for drilling, click on **Done** button from **Drilling** tab. The drilling operation will be added in **Model Tree**.

Deep

The **Deep** tool is used to drill through the workpiece. The procedure to use this tool is discussed next.

- Click on the **Deep** tool of **Pecking** drop-down from **Mill** tab; refer to Figure-51. The **Drilling** contextual tab will be displayed; refer to Figure-52.

Figure-51. Deep tol

Figure-52. Drilling tab for Deep tool

- Click on the **Tool Selection** drop-down and select the required tool for deep drilling.
- Click on the **References** tab of **Drilling** contextual tab and select the required option from **Type** drop-down.
- Click in the **Holes** selection box and select the axes of hole. You can also select multiple holes with the use of **CTRL** key.
- Click on the **Parameters** tab and specify the required parameters for drilling.
- Click on the **Clearance** tab and select the retract plane for workpiece as discussed earlier.
- After specifying the parameters, click on **Done** button from **Drilling** contextual tab.

Breakchip

The procedure to use this tool is discussed next.

- Click on the **Breakchip** tool from **Pecking** drop-down from **Mill** tab; refer to Figure-53. The **Drilling** contextual tab will be displayed.

Figure-53. Breakchip tool

- Click on the Tools Selection drop-down and select the required tool for drilling.
- Click on the **References** tab and select the axes for holes.
- Click on the **Parameters** tab and specify the required parameters for drilling.
- After specifying the parameters, click on the **Done** button from **Drilling** tab. The **Drilling** operation will be added in Model Tree.

Similarly, you can use other tools of **Holemaking Cycles** Panel.

PRACTICAL 1

In this practical, we will create toolpath for the given model; refer to Figure-54.

Figure-54. Practical 11

The first step before we start machining is to identify the operation that are required to machine the part. We can identify from the part that there is turning and drilling operations required. Note that these basic machining operations can be sub-divided into multiple operations based on the capabilities of the CAM software. We will start one by one various operations to be performed.

Starting the NC Assembly

- Double click on **Creo Parametric 4.0** icon from your computer's desktop. The **Creo Parametric 4.0** application window will be displayed; refer to Figure-55.

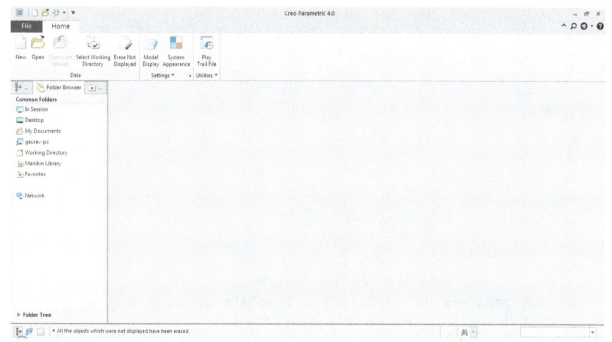

Figure-55. Creo Paramteric 4.0 starting window

* Click on the **New** button from **Ribbon**. The **New** dialog box will be displayed; refer to Figure-56.

Figure-56. New dialog box

* Click on **Manufacturing** radio button from **Type** area and **NC Assembly** radio button from **Sub-Type** area.
* Click on the **Name** edit box and enter **Practical 1** as the name of file.
* Clear the **Use default template** check box from **New** dialog box.
* Click on **OK** button from **New** dialog box. The **New File Options** dialog box will be displayed; refer to Figure-57.

Figure-57. New File Options dialog box

* Click on **mmns mfg nc** option from **Template** section and specify the required details of **Parameters** section.
* Click on **OK** button from **New File Options** dialog box. The **NC Assembly** application window will be displayed; refer to Figure-58.

Figure-58. The NC Assembly application window

Adding Reference Model

- Click on **Assemble Reference Model** button of **Reference Model** drop-down from **Manufacturing** tab. The **Open** dialog box will be displayed; refer to Figure-59.

Figure-59. Open dialog box

- Select **Practical 1** file from resource folder and click on **Open** button. The **Component Placement** tab will be displayed along with the reference model; refer to Figure-60.

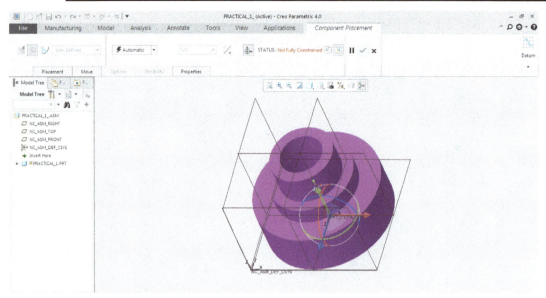

Figure-60. Reference model of Practical 1

- Click on **Default** button from **Constrain** drop-down and the **STATUS** will be changed from **Not Fully Constrained** to **Fully Constrained**.
- Click on **Done** button, the **NC Assembly** application window will be displayed.

Creating Workpiece

- Click on the **Automatic Workpiece** button of **Workpiece** drop-down from **Manufacturing** tab. The **Auto Workpiece creation** tab will be displayed; refer to Figure-61.

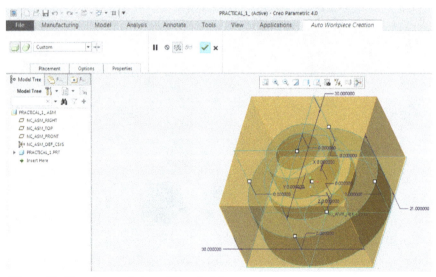

Figure-61. Creating workpiece of Practical 1

- Click on **Create Round Workpiece** button from **Auto Workpiece Creation** tab.
- Click on **Options** tab of **Auto Workpiece Creation** tab.
- Click on **About X** edit box of **Rotation Offset** section and enter the value as **90**; Figure-62.
- Click on **Length(+)** button of **Options** tab and enter the value as **0**.
- Click on **Length(-)** button of **Options** tab and enter the value as **0**.

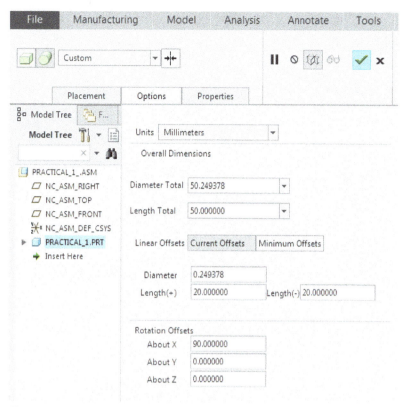

Figure-62. Options tab

- After specifying the parameters, click on **Done** button from **Auto Workpiece Creation** tab.
- The **NC Assembly** application window will be displayed along with the reference model and workpiece Figure-63.

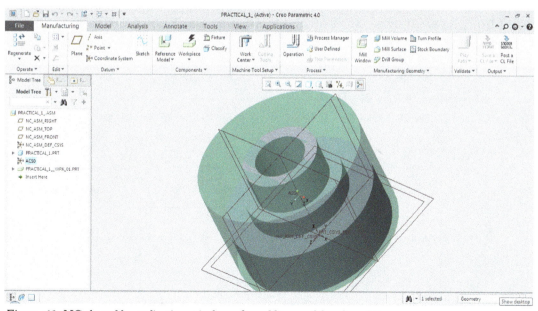

Figure-63. NC Assembly application window after adding model and workpiece

Machine Tool Setup

- Click on **Lathe** tool from **Work Center** drop-down. The **Lathe Work Center** dialog box will be displayed; refer to Figure-64.

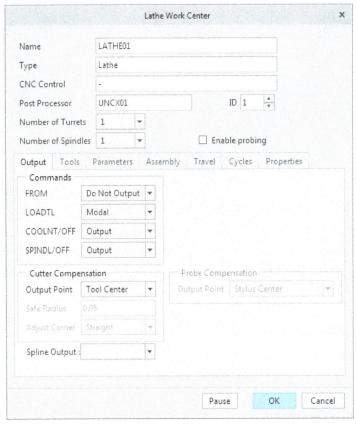

Figure-64. Lathe Work Center dialog box

- Click in the **Name** edit box and enter **TMT LATHE** as the name of machine.
- Click on **Tools** tab of **Lathe Milling Center** dialog box and click on **Turret 1** button of **Tools** tab. The **Tools Setup** dialog box will be displayed.
- Enter the parameters of tool as shown in Figure-65.

Figure-65. Tools Setup dialog box for Lathe

- After specifying the parameters, click on **Apply** button. The tool will be added in record section.
- Click on **OK** button from **Tools Setup** dialog box and **Lathe Work Center** dialog box. The Lathe operation will be added in Model Tree with the name as **TMT LATHE**.

Creating Operation

- Click on the **Operation** button of **Process** panel from **Manufacturing** tab. The **Operation** tab will be displayed; refer to Figure-66.

Figure-66. Operation tab

- Click on coordinate system from model to set the program zero point and click on **Done** button from **Operation** tab. The **Turn** tab will be added after **Applications** tab; refer to Figure-67.

Figure-67. Turn tab

Creating Turn Profile

For lathe machine you need to create a turn profile. The procedure to create profile is discussed next.

- Click on the **Turn Profile** button of **Manufacturing Geometry** panel from **Manufacturing** tab. The **Turn Profile** tab will be displayed; refer to Figure-68.

Figure-68. The Turn Profile tab

- Click on **Sketch** button which is highlighted in above figure. The **Sketch** tab will be displayed.
- You need to project the model as displayed in Figure-69.

Figure-69. Projecting the sketch

- After projecting, click on **OK** button from **Sketch** tab. The Turn Profile tab will be displayed; refer to Figure-70.

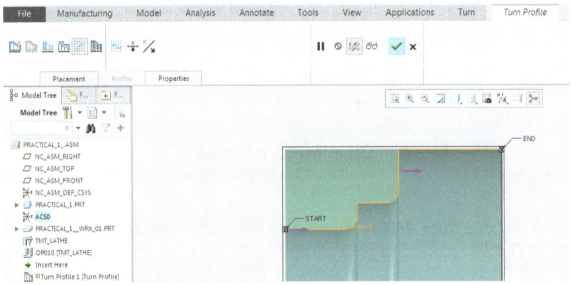

Figure-70. Turn Profile tab along with projected model

- Click on **Done** button from **Turn Profile** tab. The Turn Profile operation will be added in Model Tree.

Creating Turning Operation

- Click on **Area Turning** tool of **Turning** panel from **Turn** tab. The **Area Turning** tab will be displayed; refer to Figure-71.

Figure-71. Area Turning tab

- Click on **Tools selection** drop-down and select the **T0001** tool for turning.
- Click on **Parameters** tab and specify the parameters as shown in Figure-72.

Figure-72. Parameters Tab of area turning

- Click on **Area Turning Cut** tool from **Tool Motions** tab. The **Area Turning Cut** dialog box will be displayed; refer to Figure-73.

Figure-73. Area Turning Cut dialog box

- Click in **Turn Profile** selection box of **Area Turning Cut** dialog box and select the recently created turn profile; refer to Figure-74. Preview of toolpath will be displayed; refer to Figure-75.

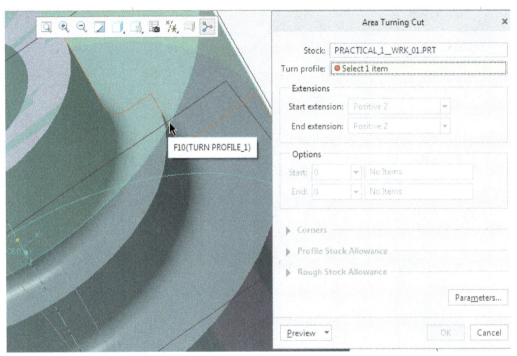

Figure-74. Selection of turn profile

Figure-75. Selection for turn profile

- Click on **OK** button from **Area Turning Cut** dialog box and **Done** button from **Area Turning** tab. The Area Turning operation will be added in Model Tree as **Area Turning 1**.

Creating Drilling Operation

- Click on **Standard** tool of **Holemaking Cycles** panel from **Turn** tab. The Drilling contextual tab will be displayed.
- Click on Tools Selection drop-down from **Drilling** tab and click on **Edit tools** button. The **Tools Setup** dialog box will be displayed.
- Create the Drill and Boring tool with parameters as shown in Figure-76.

Figure-76. Drill and Boring tool

- After creating the tools, click on **OK** button from **Tools Setup** dialog box. Both the tools will be created and added in tools selection list.
- Click on **References** tab from **Drilling** tab and select **Axes** from **Type** drop-down.
- Click on **Holes** selection box and select the axes of hole from model; refer to Figure-77.

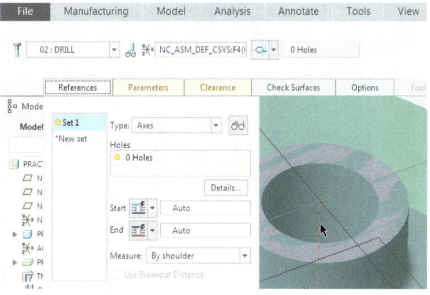

Figure-77. Selectin of axes

- Click on **Parameters** tab and specify the parameters as shown in Figure-78.

Figure-78. Parameters Tab of Drilling

- Click on **Reference** selection box of **Clearance** tab from **Drilling** tab and select the face of workpiece as shown in Figure-79.

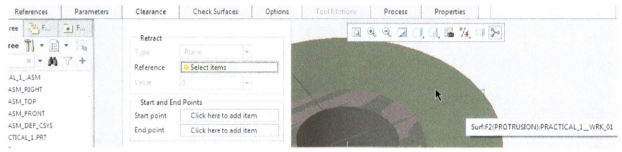

Figure-79. Selection of retract plane for drilling

- Click on **Value** edit box of **Retract** section and enter the value as 4.
- After specifying the various parameters, click on **Done** button from **Drilling** tab. The drilling operation will be added in Model Tree with the name as Drilling 1.

Creating Boring Operation

- Click on the **Boring** tool of **Holemaking Cycles** panel from **Turn** tab. The **Boring** contextual tab will be displayed; refer to Figure-80.

Figure-80. Boring Tab

- Click on the **Tools Selection** drop-down and select the **Boring** tool.
- Click on **References** tab from **Boring** tab and select **Axes** from **Type** drop-down.
- Click on **Holes** selection box and select the axes of hole from model which were used earlier for drilling hole.
- Click on **Parameters** tab and specify the parameters as shown in Figure-78.

Figure-81. Parameters Tab of Drilling

- After specifying the parameters, click on **Done** button from **Boring** contextual tab. The **Boring** operation will be added in Model Tree with the name as Boring 1.

Simulating Material Removal

The process of creating part from workpiece is competed. Now, you can simulate the cutting material from workpiece by using the **Material Removal Simulation** tool and the procedure is discussed next.

- Right click on OPO10[TECHNO] or operation name. A shortcut menu will be displayed; refer to Figure-82.

Figure-82. Material Simulation button

- Click on **Material Removal Simulation** button from menu. The **Material Removal** tab will be displayed; refer to Figure-83.

Figure-83. Material removal tab

- Click on **Start the Simulation Player** button from **Material Removal** tab. The **Play Simulation** dialog box will be displayed; refer to Figure-84.

Figure-84. Play Simulation dialog box

- Now, adjust the speed as required from **Display Speed** section and click on **Play** button. The animation will start; refer to Figure-85.

Figure-85. Animation of tool cutting

PRACTICAL 2

In this practical we will create toolpath for the given model; refer to Figure-86.

Figure-86. Practical 2

The first step before we start machining is to identify the operation that are required to machine the part. We can identify from the part that there is milling and drilling operations required. Note that these basic machining operations can be sub-divided into multiple operations based on the capabilities of the CAM software. We will start one by one various operations to be performed.

Start the **NC Assembly** application window as discussed earlier.

Adding Reference Model

• Click on **Assemble Reference Model** button of **Reference Model** drop-down from **Manufacturing** tab. The **Open** dialog box will be displayed; refer to Figure-87.

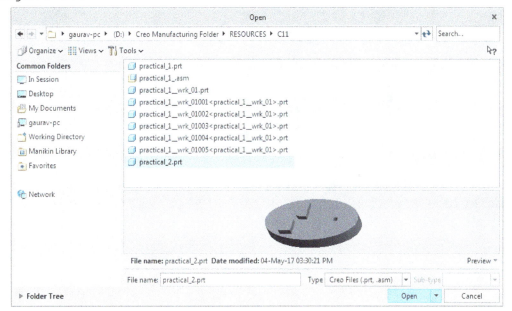

Figure-87. The Open dialog box2

- Select the **Practical 2** part file and click on **Open** button. The **Component Placement** tab will be displayed along with the model; refer to Figure-88.

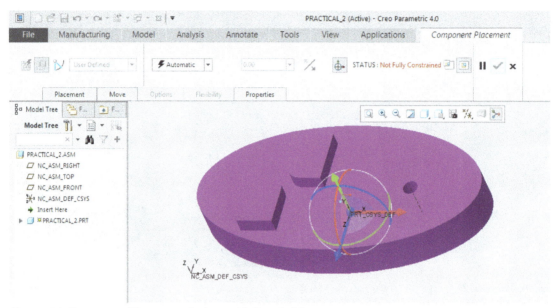

Figure-88. Component Placement tab along with Practical 2

- Select **Default** button from **Constrain** drop-down of **Component Placement** tab and click on **Done** button. The **Manufacturing** tab will be displayed along with reference model; refer to Figure-89.

Figure-89. Manufacturing tab with model

Creating Workpiece

- Click on the **Automatic Workpiece** tool from **Workpiece** drop-down. The **Auto Workpiece Creation** tab will be displayed; refer to Figure-90.

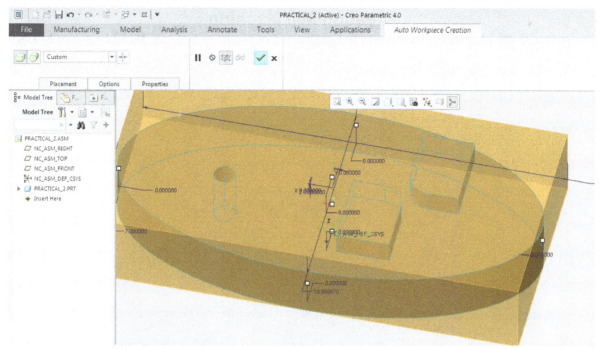

Figure-90. Creating workpiece for Practical 2

- Click on **Done** button from **Auto Workpiece Creation** tab. The model will be displayed along with workpiece.

Creating Mill Window

- Click on **Mill Window** button of **Manufacturing Geometry** panel from **Manufacturing** tab. The **Mill Window** tab will be displayed; refer to Figure-91.

Figure-91. Mill Window tab

- Click in **Window Plane** selection box of **Placement** tab and click on top face of workpiece as shown in Figure-92.

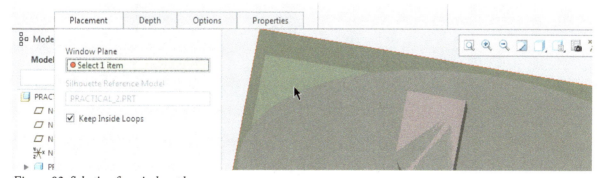

Figure-92. Selection for window plane

- Click on **Done** button from **Mill Window** tab. The **NC Assembly** application window will displayed.

Creating Mill Machine

• Click on **Mill** tool from **Work Center** drop-down. The **Milling Work Center** dialog box will be displayed; refer to Figure-93.

Figure-93. Milling Work Center dialog box

• Click on **Tools** button of **Tools** tab. The **Tools Setup** dialog box will be displayed.
• Create three tools of parameters shown in Figure-94.

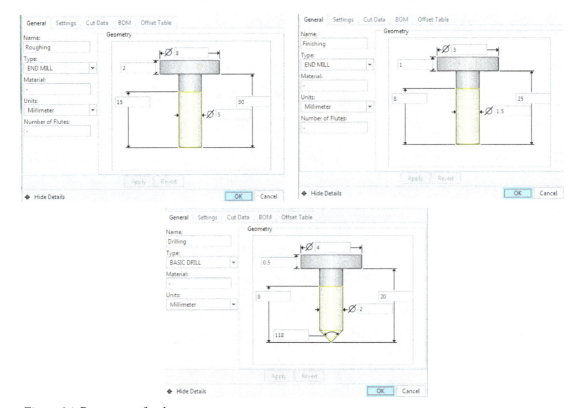

Figure-94. Parameters of tools

- After specifying the parameters, click on **Apply** button. These tools will added in records section of **Tools Setup** dialog box; refer to Figure-95.

Figure-95. Records section of Tools Setup dialog box

- Click on **OK** button from **Tools Setup** dialog box and **Milling Work Center** dialog box. The **Mill** machine will be added in Model Tree.

Creating Operation

- Click on **Operation** button of **Process** Panel from **Manufacturing** tab. The **Operation** tab will be displayed.
- Click on **Coordinate System** button from **Datum** drop-down. The **Coordinate system** dialog box will be displayed.
- Select the face of workpiece as shown in Figure-96.

Figure-96. placing coordinate system2

- Pick the one green handle of coordinate system and place it to nearby edge. Similarly repeat with second green handle. The **Origin** tab will updated.

- Click on **Offset** drop-down of **Offset references** section and select **Allign** button; refer to Figure-97. The coordinate system will aligned correctly; refer to Figure-98.

Figure-97. Setting up coordinate system

Figure-98. Correctly allignes coordinate System

- Click on **OK** button from **Coordinate System** dialog box and **Resume** button from **Operation** tab.
- The coordinate system will selected automatically. Click on **Done** button from **Operation** tab.

Creating Mill Machine Operations

Roughing

- Click on **Roughing** tool of **Milling** panel from **Mill** tab. The **Roughing** tab will be displayed; refer to Figure-99.

Figure-99. Roughing Tab

- Click on **Tools Selection** drop-down and select **Roughing** tool.
- Click on **Mill Window** selection box from **References** tab and select the recently mill window created earlier.
- Click on **Parameters** tab and specify the required parameters as shown; refer to Figure-100.

Figure-100. Parameters for roughing

- Click on reference selection box of **Clearance** tab from **Roughing** tab and select the upper face of workpiece.
- Click on the **Value** edit box of **Retract** section and enter the value as 4.
- After specifying the parameters, click on **Done** button from **Roughing** tab. The roughing operation will be added in Model Tree with the name as **Roughing 1**.

Finishing

- Click on **Finishing** tool of **Milling** panel from **Manufacturing** tab. The **Finishing** tab will be displayed; refer to Figure-101.

Figure-101. Finishing Tab

- Select **Mill Window** in references tab as discussed earlier.
- Click on **Parameters** tab and specify the required parameters, refer to Figure-102.

Figure-102. Paramters for finishing

- Click on **Done** button from **Finishing** tab. The **Finishing** operation will be added in Model Tree with the name as **Finishing 1**.

Drilling

- Click on **Standard** tool of **Holemaking Cycles** panel from **Mill** tab. The **Drilling** tab will be displayed; refer to Figure-103.

Figure-103. Drilling tab

- Click in **Holes** selection box of **References** tab from **Drilling** tab and select the axes of hole from model; refer to Figure-104.

Figure-104. Selection of hole

- Click on **Parameters** tab of **Drilling** tab and specify the parameters as shown in Figure-105.

References	Parameters	Clearance	Check Surfaces

CUT_FEED	3
FREE_FEED	-
TOLERANCE	0.01
BREAKOUT_DISTANCE	0
SCAN_TYPE	SHORTEST
CLEAR_DIST	2
PULLOUT_DIST	-
SPINDLE_SPEED	6500
COOLANT_OPTION	OFF

Figure-105. Parameters for finishing

- After specifying the parameters, click on **Done** button from **Drilling** tab. The **Drilling** operation will be added in Model Tree with the name as **Drilling 1**.

If you want to simulate the material removal based on created operations then repeat the simulation steps as discussed in previous practical.

PRACTICE 1
Create the Mill machining process and toolpath of the given Figure-106.

Figure-106. Practice 1

PRACTICE 2

Create the Lathe machining process and tool path of the given Figure-107.

Figure-107. Practie 2

Index

Ethics of an Engineer

- Engineers shall hold paramount the safety, health and welfare of the public and shall strive to comply with the principles of sustainable development in the performance of their professional duties.

- Engineers shall perform services only in areas of their competence.

- Engineers shall issue public statements only in an objective and truthful manner.

- Engineers shall act in professional manners for each employer or client as faithful agents or trustees, and shall avoid conflicts of interest.

- Engineers shall build their professional reputation on the merit of their services and shall not compete unfairly with others.

- Engineers shall act in such a manner as to uphold and enhance the honor, integrity, and dignity of the engineering profession and shall act with zero-tolerance for bribery, fraud, and corruption.

- Engineers shall continue their professional development throughout their careers, and shall provide opportunities for the professional development of those engineers under their supervision.

www.ingramcontent.com/pod-product-compliance
Lightning Source LLC
Chambersburg PA
CBHW080147060326
40689CB00018B/3877